Lecture Notes in Physics

For information about Vols. 1–67, please contact your bookseller or Springer-Verlag.

Vol. 68: Y. V. Venkatesh, Energy Methods in Time-Varying System Stability and Instability Analyses. XII, 256 pages. 1977.

Vol. 69: K. Rohlfs, Lectures on Density Wave Theory. VI, 184 pages. 1977.

Vol. 70: Wave Propagation and Underwater Acoustics. Edited by J. Keller and J. Papadakis. VIII. 287 pages. 1977.

Vol. 71: Problems of Stellar Convection. Proceedings 1976. Edited by E. A. Spiegel and J. P. Zahn. VIII, 363 pages. 1977.

Vol. 72: Les instabilités hydrodynamiques en convection libre forcée et mixte. Edité par J. C. Legros et J. K. Platten. X, 202 pages. 1978.

Vol. 73: Invariant Wave Equations. Proceedings 1977. Edited by G. Velo and A. S. Wightman. VI, 416 pages. 1978.

Vol. 74: P. Collet and J.-P. Eckmann, A Renormalization Group Analysis of the Hierarchical Model in Statistical Mechanics. IV, 199 pages. 1978.

Vol. 75: Structure and Mechanisms of Turbulence I. Proceedings 1977. Edited by H. Fiedler. XX, 295 pages. 1978.

Vol. 76: Structure and Mechanisms of Turbulence II. Proceedings 1977. Edited by H. Fiedler. XX, 406 pages. 1978.

Vol. 77: Topics in Quantum Field Theory and Gauge Theories. Proceedings, Salamanca 1977. Edited by J. A. de Azcárraga. X, 378 pages 1978.

Vol. 78: Böhm, The Rigged Hilbert Space and Quantum Mechanics. IX, 70 pages. 1978.

Vol. 79: Group Theoretical Methods in Physics. Proceedings, 1977. Edited by P. Kramer and A. Rieckers. XVIII, 546 pages. 1978.

Vol. 80: Mathematical Problems in Theoretical Physics. Proceedings, 1977. Edited by G. Dell'Antonio, S. Doplicher and G. Jona-Lasinio. VI, 438 pages. 1978.

Vol. 81: MacGregor, The Nature of the Elementary Particle. XXII, 482 pages. 1978.

Vol. 82: Few Body Systems and Nuclear Forces I. Proceedings, 1978. Edited by H. Zingl, M. Haftel and H. Zankel. XIX, 442 pages. 1978.

Vol. 83: Experimental Methods in Heavy Ion Physics. Edited by K. Bethge. V, 251 pages. 1978.

Vol. 84: Stochastic Processes in Nonequilibrium Systems, Proceedings, 1978. Edited by L. Garrido, P. Seglar and P. J. Shepherd. XI, 355 pages. 1978

Vol. 85: Applied Inverse Problems. Edited by P. C. Sabatier. V, 425 pages. 1978.

Vol. 86: Few Body Systems and Electromagnetic Interaction. Proceedings 1978. Edited by C. Ciofi degli Atti and E. De Sanctis. VI, 352 pages. 1978.

Vol. 87: Few Body Systems and Nuclear Forces II, Proceedings, 1978. Edited by H. Zingl, M. Haftel, and H. Zankel. X, 545 pages. 1978.

Vol. 88: K. Hutter and A. A. F. van de Ven, Field Matter Interactions in Thermoelastic Solids. VIII, 231 pages. 1978.

Vol. 89: Microscopic Optical Potentials, Proceedings, 1978. Edited by H. V. von Geramb. XI, 481 pages. 1979.

Vol. 90: Sixth International Conference on Numerical Methods in Fluid Dynamics. Proceedings, 1978. Edited by H. Cabannes, M. Holt and V. Rusanov. VIII, 620 pages. 1979.

Vol. 91: Computing Methods in Applied Sciences and Engineering, 1977, II. Proceedings, 1977. Edited by R. Glowinski and J. L. Lions. VI, 359 pages. 1979.

Vol. 92: Nuclear Interactions. Proceedings, 1978. Edited by B. A. Robson. XXIV, 507 pages. 1979.

Vol. 93: Stochastic Behavior in Classical and Quantum Hamiltonian Systems. Proceedings, 1977. Edited by G. Casati and J. Ford. VI, 375 pages. 1979.

Vol. 94: Group Theoretical Methods in Physics. Proceedings, 1978. Edited by W. Beiglböck, A. Böhm and E. Takasugi. XIII, 540 pages. 1979.

Vol. 95: Quasi One-Dimensional Conductors I. Proceedings, 1978. Edited by S. Barišić, A. Bjeliš, J. R. Cooper and B. Leontić. X, 371 pages. 1979.

Vol. 96: Quasi One-Dimensional Conductors II. Proceedings 1978. Edited by S. Barišić, A. Bjeliš, J. R. Cooper and B. Leontić. XII, 461 pages. 1979.

Vol. 97: Hughston, Twistors and Particles. VIII, 153 pages. 1979.

Vol. 98: Nonlinear Problems in Theoretical Physics. Proceedings, 1978. Edited by A. F. Rañada. X, 216 pages. 1979.

Vol. 99: M. Drieschner, Voraussage – Wahrscheinlichkeit – Objekt. XI, 308 Seiten. 1979.

Vol. 100: Einstein Symposion Berlin. Proceedings 1979. Edited by H. Nelkowski et al. VIII, 550 pages. 1979.

Vol. 101: A. Martin-Löf, Statistical Mechanics and the Foundations of Thermodynamics. V, 120 pages. 1979.

Vol. 102: H. Hora, Nonlinear Plasma Dynamics at Laser Irradiation. VIII, 242 pages. 1979.

Vol. 103: P. A. Martin, Modèles en Mécanique Statistique des Processus Irréversibles. IV, 134 pages. 1979.

Vol. 104: Dynamical Critical Phenomena and Related Topics. Proceedings, 1979. Edited by Ch. P. Enz. XII, 390 pages. 1979.

Vol. 105: Dynamics and Instability of Fluid Interfaces. Proceedings, 1978. Edited by T. S. Sørensen. V, 315 pages. 1979.

Vol. 106: Feynman Path Integrals, Proceedings, 1978. Edited by S. Albeverio et al. XI, 451 pages. 1979.

Vol. 107: J. Kijowski, W. M. Tulczyjew, A Symplectic Framework for Field Theories. IV, 257 pages. 1979.

Vol. 108: Nuclear Physics with Electromagnetic Interactions. Proceedings, 1979. Edited by H. Arenhövel and D. Drechsel. IX, 509 pages. 1979.

Vol. 109: Physics of the Expanding Universe. Proceedings, 1978. Edited by M. Demiański. V, 210 pages. 1979.

Vol. 110: D. A. Park, Classical Dynamics and Its Quantum Analogues. VIII, 339 pages. 1979.

Vol. 111: H.-J. Schmidt, Axiomatic Characterization of Physical Geometry. V, 163 pages. 1979.

Vol. 112: Imaging Processes and Coherence in Physics. Proceedings, 1979. Edited by M. Schlenker et al. XIX, 577 pages. 1980.

Vol. 113: Recent Advances in the Quantum Theory of Polymers. Proceedings 1979. Edited by J.-M. André et al. V, 306 pages. 1980.

Lecture Notes in Physics

151

Integrable Quantum Field Theories

Proceedings of the Symposium
Held at Tvärminne, Finland, 23–27 March, 1981

Edited by J. Hietarinta and C. Montonen

Springer-Verlag
Berlin Heidelberg GmbH 1982

Editors

Jarmo Hietarinta
Department of Physical Sciences
University of Turku
Turku, Finland

Claus Montonen
Department of Theoretical Physics
University of Helsinki
Helsinki, Finland

ISBN 978-3-540-11190-0 ISBN 978-3-540-38976-7 (eBook)
DOI 10.1007/978-3-540-38976-7

Originally published by Springer-Verlag Berlin Heidelberg New York in 1982

2153/3140-543210

Preface

The present volume contains the invited lectures given at the Symposium on Integrable Quantum Field Theory, organised by the Research Institute for Theoretical Physics of the University of Helsinki in Tvärminne, Finland, 23-27 March, 1981.

We are grateful to Dr. E.K. Sklyanin, who, although unable to attend due to a last-minute delay, agreed to submit together with Dr. P.P. Kulish an expanded review of the work done by the Leningrad school.

The symposium received generous financial support from the Academy of Finland and NORDITA, which is gratefully acknowledged.

Jarmo Hietarinta

Claus Montonen

Contents

THE QUANTUM INVERSE METHOD AND GREEN'S FUNCTIONS FOR COMPLETELY INTEGRABLE FIELD THEORIES

H.B. Thacker

Fermi National Accelerator Laboratory

P.O. Box 500, Batavia, Illinois 60510

C o n t e n t s

I. INTRODUCTION

The development of the quantum inverse method[1-6] has provided new insight into the structure of solvable models in quantum field theory and statistical mechanics. It places the theory of completely integrable quantum systems in a unified framework and provides a powerful method for studying these systems. In this series of lectures, I will review some of these developments with particular emphasis on the study of Green's functions for integrable field theories. The approach to Green's functions which I will describe has been developed in collaboration with Dennis Creamer and David Wilkinson.[6-9] So far it has only been applied to the case of the nonlinear Schrödinger model, but it is reasonable to suspect that similar techniques can be applied to other models.

I'll begin in Section II by reviewing the direct scattering transform[1-5] by which a certain set of "scattering data" operators are defined as functionals of the local fields. I'll describe the connection between the direct transform and the more traditional Bethe's ansatz methods and briefly mention the relationship with transfer matrices in lattice statistical models. The treatment of Green's functions is built upon the inverse (Gel'fand-Levitan) transform by which the local fields are written as operator functionals of the scattering data. In Section III I'll review the derivation of the quantum Gel'fand-Levitan transform[6] for the nonlinear Schrödinger model and discuss some of its properties. Section IV sets up the general formalism for studying Green's functions via the Gel'fand-Levitan transform.[9] In Section V we'll use this formalism to study the strong coupling ($c \to \infty$) limit of the two-point function. Finally, in Section VI I review the analysis of the $c = \infty$ two-point function by Jimbo,

Miwa, Mori, and Sato[12] and show that the first two terms in a strong coupling (1/c) expansion can be expressed in closed form in terms of Painlevé functions.[9,12]

II. BETHE'S ANSATZ AND THE DIRECT SCATTERING TRANSFORM

The Nonlinear Schrödinger Model

The case we'll be considering is the nonlinear Schrödinger model, defined by the Hamiltonian

$$H = \int \left[\frac{\partial\phi^*\partial\phi}{\partial x\partial x} + c\phi^*\phi^*\phi\phi \right] dx \qquad (2.1)$$

where $\phi(x)$ is a nonrelativistic boson field with equal time commutation relations

$$[\phi(x), \phi^*(y)] = \delta(x - y) \qquad . \qquad (2.2)$$

The second term in H corresponds to a two-body delta-function potential. We'll consider the repulsive case $c > 0$, for which the problem of interest is to determine the spectrum and Green's functions for a finite density ground state $\langle\phi^*\phi\rangle \neq 0$. This is analogous to the problem faced in relativistic models like sine Gordon/massive Thirring, where the physical vacuum is a many-body Bethe's ansatz state.

Before introducing the quantum inverse method, I'll review the Bethe ansatz approach to (2.1). In this approach we write down many-body states

$$|\Psi\rangle = \int dx_1 \ldots dx_N \psi(x_1 \ldots x_N)\, \phi^*(x_1) \ldots \phi^*(x_N)\, |0\rangle \qquad (2.3)$$

and try to choose the wave function ψ so that $|\Psi\rangle$ is an exact eigenstate of H. The correct wave functions ψ have the characteristic Bethe ansatz form which I'll now describe. Consider first the two-particle state,

$$|\Psi(k_1, k_2)\rangle = \int dx_1 dx_2 e^{i(k_1x_1+k_2x_2)} \left\{ \theta(x_1 - x_2) + S(k_{21})\theta(x_2 - x_1) \right\} \phi^*(x_1)\phi^*(x_2)|0\rangle \tag{2.4}$$

where $k_{21} \equiv k_2 - k_1$ and

$$S(k) = \frac{k - ic}{k + ic} \tag{2.5}$$

is the two-body phase shift. The fact that (2.4) is an eigenstate of H may be shown directly by applying the operator H and using integration by parts to bring the kinetic energy derivative $-\partial^2/\partial x^2$ onto the two-body wave function. This gives

$$H|\Psi(k_1, k_2)\rangle = (k_1^2 + k_2^2)|\Psi(k_1, k_2)\rangle \quad . \tag{2.6}$$

In the derivation of this result, there is a leftover term proportional to $\delta(x_1 - x_2)$ coming from kinetic energy derivatives acting on the step functions in (2.4). This term is exactly cancelled by the δ-function interaction term. We can also write the two-body state in a different form by changing the normalization

$$|\Phi(k_1k_2)\rangle \equiv \left(1 + \frac{ic}{k_{21}}\right)|\Psi(k_1k_2)\rangle$$

$$= \int dx_1 dx_2 e^{i(k_1x_1+k_2x_2)} \left\{ 1 - \frac{ic}{k_{21}} \varepsilon(x_{21}) \right\} \phi^*(x_1)\phi^*(x_2)|0\rangle \tag{2.7}$$

where $x_{21} \equiv x_2 - x_1$.

The Bethe ansatz[13] for this model consists of a generalization of (2.4) to an arbitrary number of particles N such that

$$H|\Psi(k_1 \ldots k_N)\rangle = \left(\sum_{i=1}^{N} k_i^2 \right) |\Psi(k_1 \ldots k_N)\rangle \quad . \tag{2.8}$$

Just as in the two-body case (2.4) the wavefunction can be written as a sum over N! orderings of the coordinates $x_{P_1} > x_{P_2} > \ldots > x_{P_N}$ where $(P_1, P_2, \ldots, P_N)$ is some permutation of $(1, 2, \ldots, N)$,

$$|\Psi(k_1 \ldots k_N)\rangle = \int dx_1 \ldots dx_N \exp\left(i \sum_1^N k_i x_i\right) \left\{ \sum_{P \in S_N} \theta\left(x_{P_1} > \ldots > x_{P_N}\right) \right.$$

$$\left. \times \prod_{\substack{i<j \\ P_i > P_j}} S\left(k_{P_i P_j}\right) \right\} \phi^*(x_1) \ldots \phi^*(x_N)|0\rangle . \tag{2.9a}$$

The unnormalized eigenstates analogous to (2.7) are written

$|\Phi(k_1 \ldots k_N)\rangle =$

$$\int dx_1 \ldots dx_N \exp\left(i \sum_1^N k_i x_i\right) \prod_{i<j} \left\{ 1 - \frac{ic}{k_i - k_j} \varepsilon(x_i - x_j) \right\} \phi^*(x_1) \ldots \phi^*(x_N)|0\rangle \tag{2.9b}$$

Spectral properties of finite density states

The finite density system is traditionally studied by placing an N-body system in a periodic box of length L and letting $N \to \infty$ with N/L = density fixed. The wave function

$$\psi(x_1 \ldots x_N) = \langle 0|\phi(x_1) \ldots \phi(x_N)|\Phi(k_1 \ldots k_N)\rangle \tag{2.10}$$

is required to satisfy periodic boundary conditions (PBC's)

$$\psi(-L/2, x_2 \ldots x_N) = \psi(L/2, x_2 \ldots x_N) \tag{2.11}$$

which gives

$$e^{ik_i L} = \prod_{j \neq i} S(k_{ji}) \quad . \tag{2.12}$$

It is convenient to take the log of the PBC's

$$k_i L = \sum_{j \neq i} \theta(k_j - k_i) + 2\pi n_i \tag{2.13}$$

where

$$\theta(k) = -i \log S(k) \quad . \tag{2.14}$$

The choice of n_i's in (2.13) is related to the choice of branch for the log in (2.14). The physical phase shift has a discontinuity of 2π at $k = 0$. This phase shift vanishes as $c \to 0$ and the description of the ground state of the system is bosonic, i.e. $n_i = 0$ for all i. Instead it is convenient and conventional to choose the phase shift which is continuous at $k = 0$ for finite c and becomes a step function as $c \to 0$. For this choice, the ground state has a fermionic description, $n_{i+1} - n_i = 1$.

By subtracting adjacent PBC's in the ground state we obtain

$$k_{i+1} - k_i = \frac{1}{L} \sum_j \left[\theta(k_j - k_{i+1}) - \theta(k_j - k_i)\right] + \frac{2\pi}{L} \, . \quad (2.15)$$

As $L \to \infty$, $N \to \infty$, N/L fixed

$$\rho(k_i) = \frac{1}{L(k_{i+1} - k_i)} \to \text{continuous function} \equiv \rho(k) \quad (2.16)$$

and the PBC's (2.15) reduce to an integral equation for the ground state density function

$$2\pi\rho(k) = 1 + \int_{-k_F}^{k_F} \Delta(k - k')\rho(k')dk' \quad (2.17)$$

where

$$\Delta(k) = \frac{\partial\theta(k)}{\partial k} = \frac{2c}{k^2 + c^2} \quad . \quad (2.18)$$

The ground state is a Fermi sea of closely packed modes between $-k_F$ and k_F. Excited states are formed by removing modes from the sea and placing them above the surface, forming particle-hole pairs. The particle-hole spectrum was first worked out by Lieb.[13] In the formulation of Yang and Yang,[14] the spectrum is given by a single function $\varepsilon(k)$ which satisfies a linear integral equation

$$\varepsilon(k) = k^2 - \mu + \int_{-k_F}^{k_F} \Delta(k - k')\varepsilon(k') \frac{dk'}{2\pi} \quad (2.19)$$

where μ is fixed by the requirement $\varepsilon(\pm k_F) = 0$. The excitation energy of a particle at k_p and a hole at k_h is given by

$$E = \varepsilon(k_p) - \varepsilon(k_h) \quad . \tag{2.20}$$

In the Yang and Yang formulation, there is a similar excitation function $\varepsilon(k)$ at finite temperature, which satisfies a nonlinear equation

$$\varepsilon(k) = k^2 - \mu - \frac{1}{\beta}\int \Delta(k - k')\log\,(1 + e^{-\beta\varepsilon(k')})\,\frac{dk'}{2\pi} \tag{2.21}$$

where μ = chemical potential and $\beta = 1/kT$. This reduces to (2.19) for $\beta \to \infty$. The function $\varepsilon(k)$ also determines the equilibrium thermodynamics, e.g. the pressure of a gas as a function of β and μ is

$$P = \frac{1}{\beta}\int \frac{dk}{2\pi}\log\,(1 + e^{-\beta\varepsilon(k)}) \quad . \tag{2.22}$$

The function $\varepsilon(k)$ is of central importance in the model. (Similar functions can be constructed for other models, e.g. massive Thirring/sine Gordon.) It will reappear in the theory of Green's functions.

Quantum Inverse Method

In the classical inverse scattering method,[15-16] we solve the initial value problem for a nonlinear field equation by considering a linear "Lax pair":

$$\frac{\partial}{\partial x^\mu}\Psi(x, \zeta) = iQ_\mu(x, \zeta)\Psi(x, \zeta) \quad . \tag{2.23}$$

In the simplest applications, $Q_\mu(x, \zeta)$ is a 2×2 matrix which depends on the local field $\phi(x, t)$, and on an eigenvalue ζ. If we think of the spatial component of the Lax pair as a time independent eigenvalue (scattering) problem, the local field $\phi(x)$ plays the role of a scattering potential, and (2.23) defines a one-to-one mapping between the field $\phi(x)$ at a fixed time t and the scattering data associated with the linear eigenvalue problem. The key point is that, by judicious choice of the matrices $Q_\mu(x, \zeta)$ we may interpret the original nonlinear equation of motion as the consistency (integrability) condition obtained by cross-differentiation of the Lax pair, which gives

$$F_{\mu\nu} = \partial_\nu Q_\mu - \partial_\mu Q_\nu + i\left[Q_\mu, Q_\nu\right] = 0 \quad . \tag{2.24}$$

With the particular choice

$$Q_0 = \begin{pmatrix} \frac{k^2}{2} - c\phi^*\phi & \sqrt{c}(k\phi + i\phi_x) \\ -\sqrt{c}(k\phi^* - i\phi_x^*) & \frac{k^2}{2} + c\phi^*\phi \end{pmatrix} \tag{2.25}$$

$$Q_1 = \begin{pmatrix} \frac{k}{2} & \sqrt{c}\phi \\ -\sqrt{c}\phi^* & -\frac{k}{2} \end{pmatrix} \tag{2.26}$$

then $F_{\mu\nu} = 0$ becomes the nonlinear Schrödinger equation

$$i\partial_0\phi = -\partial_1{}^2\phi + c|\phi|^2\phi \quad . \tag{2.27}$$

From this result it follows that the scattering data a(k), b(k) (where 1/a = transmission coefficient and b/a = reflection coefficient) have a trivial time dependence

$$a(k,\ t) = a(k,0) \tag{2.28a}$$

$$b(k,\ t) = e^{-ik^2t}b(k,0) \quad . \tag{2.28b}$$

The inverse method solves the initial value problem much like Fourier transformation is used to solve a linear problem. The direct transform maps $\phi(x) \to a(k), b(k)$ at time $t = 0$. The time evolution of a and b from $t = 0$ to some later time t is given by (2.28). At time t we must perform an inverse transform which maps $a(k,\ t)$, $b(k,\ t)$ back into the field configuration $\phi(x,\ t)$. This last step is accomplished by the Gel'fand-Levitan equation.

In this section I'll discuss the quantum generalization of the direct transform, the significance of $a(k)$ and $b(k)$ as quantum operators, and the relationship with Bethe's ansatz. In the following section, I'll discuss the generalization of the Gel'fand-Levitan (inverse) transform, which is the centerpiece for the treatment of Green's functions in the remaining sections.

The quantum inverse method for the nonlinear Schrödinger model is based on a normal ordered operator version of the Zakharov-Shabat eigenvalue problem (2.23)

$$\frac{\partial}{\partial x}\Psi(x,\ k) = i :Q_1(x,\ k)\Psi(x,\ k): \tag{2.29}$$

A particular solution is specified by choosing a boundary condition. Requiring $\Psi(x_0,\ k) = I$ = identity matrix, we can write the solution to (2.29) formally as a path ordered exponential,

$$\Psi(x,\ k) \quad = \quad :\ P\ \exp\ i \int_{x_0}^{x} Q_1(y,\ k)dy\ : \qquad . \tag{2.30}$$

The solution $\Psi(x,\ k)$ is a nonlocal string functional of the field operator $\phi(x)$. If $\phi(x) \rightarrow 0$ weakly as $|x| \rightarrow \pm\infty$ we see that

$$\Psi(x,\ k) \xrightarrow[|x|\rightarrow\pm\infty]{} V(x,\ k) \times (\text{constant matrix}) \tag{2.31}$$

where

$$V(x,\ k) \quad = \quad \begin{pmatrix} e^{ik\ x/2} & 0 \\ 0 & e^{-ik\ x/2} \end{pmatrix} \qquad . \tag{2.32}$$

The scattering data operators are defined by the asymptotic form of Ψ:

$$\begin{aligned} \mathscr{T}(k) \quad &= \quad \lim_{\substack{x\rightarrow\infty \\ x_0\rightarrow-\infty}} V^{-1}(x,\ k)\Psi(x,\ k)V(x_0,\ k) \\ &= \quad \begin{pmatrix} a(k) & b^*(k) \\ b(k) & a^*(k) \end{pmatrix} \end{aligned} \tag{2.33}$$

for real k. The central result of the quantum inverse method is a set of commutation relations among the scattering data operators. This is most elegantly derived by the method of Sklyanin,[3] which is patterned after earlier work of Baxter.[17] One uses the Zakharov-Shabat equation to derive 4×4 matrix equations for the direct products $H_{12}(x) \equiv \Psi(x,\ k_1) \otimes \Psi(x,\ k_2)$ and $H_{21}(x) = \Psi(x,\ k_2) \otimes \Psi(x,\ k_1)$. We get

$$\frac{\partial}{\partial x} H_{12} \quad = \quad i\ :\ \Gamma_{12}H_{12}\ : \tag{2.34}$$

$$\frac{\partial}{\partial x} H_{21} = i : \Gamma_{21} H_{21} : \tag{2.35}$$

where

$$\Gamma_{12} = Q(k_1) \otimes I + I \otimes Q(k_2) - ic\sigma^+ \otimes \sigma^- \quad . \tag{2.36}$$

The key observation is that the matrices Γ_{12} and Γ_{21} are equivalent under a c-number similarity transformation

$$\Gamma_{21} = \mathscr{R} \Gamma_{12} \mathscr{R}^{-1} \tag{2.37}$$

where

$$\mathscr{R} = \begin{pmatrix} 1 & 0 & 0 & 0 \\ 0 & \beta & \alpha & 0 \\ 0 & \alpha & \beta & 0 \\ 0 & 0 & 0 & 1 \end{pmatrix} \tag{2.38}$$

with

$$\alpha = \frac{k_1 - k_2}{k_1 - k_2 - ic} \qquad \beta = \frac{-ic}{k_1 - k_2 - ic} \quad . \tag{2.39}$$

This leads to the result that the direct products of the solutions are themselves related by

$$\Psi_2 \otimes \Psi_1 = R[\Psi_1 \otimes \Psi_2] R^{-1} \tag{2.40}$$

where the subscript denotes the eigenvalue.

Equation (2.40) gives a set of commutation relations among the elements of the solution matrices. At this point there are two somewhat different approaches we may follow to further investigate the model. Let me refer to these two possibilities as the finite volume approach and the infinite volume approach. In the finite volume approach we define the scattering data operators in a box by choosing $x_0 = -L/2$ in (2.29) and defining

$$\Psi(L/2,\ k) = \begin{pmatrix} A(k) & C(k) \\ B(k) & D(k) \end{pmatrix} \equiv \mathcal{T}_L(k) \tag{2.41}$$

with the commutation relations

$$\left[\mathcal{T}_L(k_2) \otimes \mathcal{T}_L(k_1)\right] \mathcal{R} = \mathcal{R} \left[\mathcal{T}_L(k_1) \otimes \mathcal{T}_L(k_2)\right] \tag{2.42}$$

where $\mathcal{R}$ is given by (2.38). By carefully taking the $L \to \infty$ limit, we obtain a somewhat simpler infinite volume algebra

$$\left[\mathcal{T}(k_2) \otimes \mathcal{T}(k_1)\right] \mathcal{R}_\infty = \mathcal{R}_\infty \left[\mathcal{T}(k_1) \otimes \mathcal{T}(k_2)\right] \tag{2.43}$$

where

$$\mathcal{R}_\infty = \begin{pmatrix} 1 & 0 & 0 & 0 \\ 0 & 0 & \gamma & 0 \\ 0 & \alpha & 0 & 0 \\ 0 & 0 & 0 & 1 \end{pmatrix} \tag{2.44}$$

$$\alpha = \frac{k_1 - k_2}{k_1 - k_2 - ic} \qquad \beta = \frac{k_1 - k_2 + ic}{k_1 - k_2} \quad . \tag{2.45}$$

In particular, we find

$$a(k)b(k') = \left(1 - \frac{ic}{k - k'}\right) b(k')a(k) \tag{2.46a}$$

$$a^*(k)b(k') = \left(1 + \frac{ic}{k - k'}\right) b(k')a^*(k) \tag{2.46b}$$

$$b^*(k)b(k') = \frac{(k - k')^2 + c^2}{(k - k')^2} b(k')b^*(k) + 2\pi a^*(k)a(k)\delta(k - k') \tag{2.46c}$$

$$[a, a^*] = [a, a] = [b, b] = 0 \quad . \tag{2.46d}$$

The commutators of a and b with the Hamiltonian may also be worked out,

$$[H, a(k)] = 0 \tag{2.47}$$

$$[H, b(k)] = k^2 b(k) \tag{2.48}$$

which is the quantum analog of (2.28). All of these results may be verified order by order using the normal ordered series expansions for a(k) and b(k):

$$a(k) = 1 + c\int dx_1 dy_1 \theta(x_1 < y_1) e^{ik(x_1 - y_1)} \phi^*(x_1)\phi(y_1) + \cdots \tag{2.49}$$

$$\frac{i}{\sqrt{c}} b(k) = \int dx_1 e^{ikx_1} \phi^*(x_1) + c\int dx_1 dx_2 dy_1 \theta(x_1 < y_1 < x_2) e^{ik(x_1 + x_2 - y_1)} \times \phi^*(x_1)\phi^*(x_2)\phi(y_1) + \cdots \tag{2.50}$$

From (2.48) we see that the states

$$|\Phi(k_1 \ldots k_N)\rangle = b(k_1)\ldots b(k_N)|0\rangle \qquad (2.51)$$

are exact eigenstates of H. From (2.50) it can be shown that the states (2.51) are precisely the unnormalized Bethe ansatz states (2.9b). The operator a(k) is diagonal on these states for all k and is the generator of an infinite number of conservation laws. In the infinite volume formalism, a particularly useful operator is the quantized reflection coefficient

$$R(k) = b(k)a^{-1}(k) \quad . \qquad (2.52)$$

This operator and its conjugate obey a simple algebra

$$R(k)R(k') = S(k' - k)R(k')R(k) \qquad (2.53)$$

$$R(k)R^{+}(k') = S(k - k')R^{+}(k')R(k) + 2\pi\delta(k - k') \qquad (2.54)$$

where

$$S(k - k') = \frac{k - k' - ic}{k - k' + ic} = \text{2-body S-matrix} \quad . \qquad (2.55)$$

States created by R^{+}'s are also eigenstates of H but with a different normalization. They are in fact the properly normalized states $|\Psi(k,\ldots k_N)\rangle$ defined in (2.9b). The R operators are of central importance in the theory of the inverse problem and Green's functions.

In the finite volume formalism it is also possible to construct Bethe's ansatz states, but this time the B-states diagonalize not A(k) but the trace of the monodromy matrix (2.41)

$$T(k) = \mathrm{Tr}\ \mathcal{T}_L(k) = A(k) + D(k) \quad . \qquad (2.56)$$

This quantity is precisely analogous to the transfer matrix in two-dimensional lattice statistics models. The states created by B's are not automatically eigenstates of T(k) as they are in the infinite volume case. Instead, a state $B(k_1)\ldots B(k_N)|0\rangle$ is an eigenstate of T(k) only if $k_1,\ldots,k_N$ satisfy periodic boundary conditions. In this approach, the PBC's follow directly from the algebra of the operators A, B, C, and D. On the other hand, in the finite volume formalism the R-operators do not have nice properties, and the Gel'fand-Levitan transform has not yet been constructed. For the remainder of these lectures we will use the infinite volume approach to study Green's functions. This will result in no loss of generality, since, as we will see, all the finite density results of Lieb and Liniger and Yang and Yang can be derived in this approach by studying finite temperature Green's functions.

Let me conclude this section with some remarks on the quantum inverse method for lattice models and its deep connection with Baxter's method for solving the eight-vertex model. This connection has been extensively developed by Faddeev and coworkers.[10,18] It leads to an elegant and general formulation of quantum integrability based on the "Yang-Baxter relation," which is a generalization of the similarity relation (2.37). Essentially, one views the Jost solutions as strings of vertices of the form

$$\Psi_m(k) = L_1(k)L_2(k)\ldots L_m(k) \tag{2.57}$$

where $L_j(k)$ is a matrix of local operators defined on lattice site j. Equation (2.57) is precisely analogous to the path-ordered exponential solution of the Zakharov-Shabat equation, Eq. (2.30). The Yang-Baxter relation is

$$\mathscr{R}\left[L_n(k) \otimes L_n(k')\right] = \left[L_n(k') \otimes L_n(k)\right]\mathscr{R} \quad . \tag{2.58}$$

For the nonlinear Schrödinger case, L_n is a 2×2 matrix of field operators and $\mathscr{R}$ is just (2.38). Equation (2.58) leads directly to the results (2.40) and (2.42). For further discussions of the Yang-Baxter relation and how it arises in various models I refer you to the literature and to the paper of Kulish and Sklyanin in these proceedings.

III. THE OPERATOR GEL'FAND-LEVITAN EQUATION

The Gel'fand-Levitan equation is a dispersion relation for a Jost solution to the Zakharov-Shabat eigenvalue problem,

$$\left(i\frac{\partial}{\partial x} + \frac{1}{2}\zeta\right)\Psi_1 = -\sqrt{c}\Psi_2\phi \tag{3.1a}$$

$$\left(i\frac{\partial}{\partial x} - \frac{1}{2}\zeta\right)\Psi_2 = \sqrt{c}\phi^*\Psi_1 \quad . \tag{3.1b}$$

Consider two column vector solutions to (3.1) defined by the boundary conditions

$$\begin{pmatrix} \psi_1 \\ \psi_2 \end{pmatrix} \xrightarrow[x\to-\infty]{} \begin{pmatrix} 1 \\ 0 \end{pmatrix} e^{i\zeta x/2} \tag{3.2}$$

$$\begin{pmatrix} \chi_1 \\ \chi_2 \end{pmatrix} \xrightarrow[x\to+\infty]{} \begin{pmatrix} 0 \\ 1 \end{pmatrix} e^{-i\zeta x/2} \tag{3.3}$$

From these boundary conditions it is easily shown that both ψ and χ admit analytic continuation into the lower half ζ-plane. Here analyticity of an operator is taken to be equivalent to analyticity of all its physical matrix elements. We will also need the conjugate solutions

$$\tilde{\psi}(x,\zeta) = \begin{pmatrix} \psi_2^*(x,\zeta^*) \\ \psi_1^*(x,\zeta^*) \end{pmatrix} \qquad \tilde{\chi} = \begin{pmatrix} \chi_2^*(x,\zeta^*) \\ \chi_1^*(x,\zeta^*) \end{pmatrix} \tag{3.4}$$

which are analytic in the upper-half ζ-plane. The Gel'fand-Levitan equation is a dispersion relation for an analytic function $\Phi(x,\zeta)$ which is constructed from these Jost solutions.

Classical case:

In the classical theory, for $\zeta = k =$ real the Jost solution ψ can be written as a linear combination of χ and $\tilde{\chi}$,

$$\psi = a\tilde{\chi} + b\chi, \tag{3.5}$$

where a and b are the scattering coefficients defined previously. Equation (3.5) may be verified by taking the Wronskian of both sides with χ and $\tilde{\chi}$ and using

$$\psi_1\chi_2 - \psi_2\chi_1 = a \tag{3.6}$$

$$\psi_2\tilde{\chi}_1 - \psi_1\tilde{\chi}_2 = b \tag{3.7}$$

Thus, along the real axis, the function $\tilde{\chi}$ which is analytic in the lower half-plane is related to the function ψa^{-1} which is analytic in the upper half-plane by

$$\psi a^{-1} = \tilde{\chi} - i\sqrt{c}R^*\chi \tag{3.8}$$

(Note: a has no zeroes in the lower half-plane for repulsive coupling $c > 0$.) Equation (3.8) suggests that we define a function

$$\Phi(x,\zeta) = \tilde{\chi}(x,\zeta)e^{-i\zeta x/2} \qquad \text{for } \mathrm{Im}\zeta > 0 \tag{3.9a}$$

$$= \psi(x,\zeta)a^{-1}(\zeta)e^{-i\zeta x/2} \qquad \text{for } \mathrm{Im}\zeta < 0 \tag{3.9b}$$

This function has a discontinuity proportional to the reflection coefficient

$$\mathrm{Disc}\,\Phi = i\sqrt{c}R^*\chi e^{-i\zeta x/2} \tag{3.10}$$

Also, from the Zakharov-Shabat equation we have

$$\Phi \to 1 \quad \text{as} \quad \zeta \to \infty \tag{3.11}$$

Thus, Φ can be reconstructed from its discontinuity,

$$\Phi(x,\zeta) = \begin{pmatrix} 1 \\ 0 \end{pmatrix} + \frac{\sqrt{c}}{2\pi}\int dk\, \frac{R^*(k)\chi(x,k)e^{-ikx/2}}{k-\zeta} \tag{3.12}$$

Evaluating just above the real axis, we obtain a coupled pair of integral equations,

$$\tilde{\chi}(x, k)e^{-ikx/2} = \begin{pmatrix} 1 \\ 0 \end{pmatrix} + \frac{\sqrt{c}}{2\pi} \int_{-\infty}^{\infty} dk \frac{R^*(k')\chi(x, k')e^{-ik'x/2}}{k' - k - i\epsilon} \tag{3.13}$$

Quantum case:

In the quantum theory, the equation (3.8) which motivated the choice (3.9) for the Φ function is not a valid operator relation. Instead we define a function

$$g(x, k) \equiv \tilde{\chi}(x, k) - i\sqrt{c}R^*(k)\chi(x, k) \tag{3.14}$$

and study the analytic continuation of g into the lower half-plane. From the Zakharov-Shabat equation, we find that g satisfies

$$(i\frac{\partial}{\partial x} + \frac{1}{2}k) g_1 = -\sqrt{c}g_2\phi \tag{3.15}$$

$$(i\frac{\partial}{\partial x} - \frac{1}{2}k) g_2 = \sqrt{c}\phi^* g_1 - ic[R^*(k), \phi^*(x)]\chi_1 \tag{3.16}$$

Note that the last term in (3.16) arises from quantum ordering. Without it we would conclude that $g = \psi a^{-1}$ as in the classical case. But the commutator $[R^*(k), \phi^*(x)]$ can be evaluated by writing $R^* = ba^{-1}$ and using Wronskian relations for b and a^{-1}. This gives

$$[R^*(k), \phi^*(x)] = (\tilde{\chi}_2 - i\sqrt{c}R^*\chi_2)\psi_2 a^{-1} \tag{3.17}$$

$$= g_2\psi_2 a^{-1}$$

Thus, the Z-S equation becomes a differential equation for $g = \begin{pmatrix} g_1 \\ g_2 \end{pmatrix}$ with coefficients which are analytic in the lower half k plane. The asymptotic form of g also has simple analytic properties. For $x \to \infty$ we have

$$g(x, k) \to \begin{pmatrix} \tilde{a}(k) \\ 0 \end{pmatrix} e^{ikx/2} \tag{3.18}$$

where

$$\tilde{a}(k) = a^*(k) - cR^*(k)a^*(k)R(k) \tag{3.19}$$

$\tilde{a}(k)$ is diagonal on the Bethe ansatz states, and we may verify that it is analytic in the lower half k-plane by studying its eigenvalues. On a one particle state we get

$$\begin{aligned} \tilde{a}(k)|k_1> &= \left[1 + \frac{ic}{k - k_1 + i\epsilon} - 2\pi c\delta(k - k_1)\right] |k_2> \\ &= \left[1 + \frac{ic}{k - k_1 - i\epsilon}\right] |k_2> \end{aligned} \tag{3.20}$$

More generally, the δ-function terms in the eigenvalue of $\tilde{a}(k)$ simply change the signs of all the $i\epsilon$'s,

$$\tilde{a}(k)| k_1 \ldots k_N> = \prod_1^N \left[1 + \frac{ic}{k - k_i - i\epsilon}\right] | k_1 \ldots k_N> \tag{3.21}$$

Thus, a function

$$\Phi(x,\zeta) = \tilde{\chi}(x,\zeta)e^{-i\zeta x/2} \qquad \text{Im}\,\zeta > 0 \tag{3.22}$$

$$= g(x,\zeta)e^{-i\zeta x/2} \qquad \text{Im}\,\zeta > 0 \tag{3.23}$$

is analytic in the full cut ζ-plane with

$$\text{Disc}\,\Phi = i\sqrt{c}R^{*}\chi \tag{3.24}$$

and

$$\Phi \sim 1 + O\left(\frac{1}{\zeta}\right) \quad \text{as} \quad \zeta \to \infty \tag{3.25}$$

This gives a pair of coupled integral equations for the operator Jost solutions χ_1 and χ_2^{*}:

$$\chi_2^{*}(x,k)e^{-ikx/2} = 1 + \frac{\sqrt{c}}{2\pi}\int_{-\infty}^{\infty} dk' \, \frac{R^{*}(k')\chi_1(x,k')e^{-ik'x/2}}{k'-k-i\epsilon} \tag{3.26a}$$

$$\chi_1(x,k)e^{ikx/2} = \frac{\sqrt{c}}{2\pi}\int_{-\infty}^{\infty} dk' \, \frac{\chi_2^{*}(x,k')R(k')e^{ik'x/2}}{k'-k+i\epsilon} \tag{3.26b}$$

Solving these integral equations (e.g. by iteration) gives χ_1 and χ_2^{*} as operator functionals of R and R^{*}:

$$\chi_1(x,k)e^{ikx/2} = -\sqrt{c}\left\{\int \frac{dk_0}{2\pi}\,\frac{e^{ik_0 x}}{k-k_0-i\epsilon}R(k_0) - c\int \frac{dp_1}{2\pi}\frac{dk_0}{2\pi}\frac{dk_1}{2\pi}\,\frac{e^{i(k_0+k_1-p_1)x}R^{*}(p_1)R(k_1)R(k_0)}{(k-k_0-i\epsilon)(p_1-k_0-i\epsilon)(p_1-k_1-i\epsilon)} + \dots\right\} \tag{3.27}$$

$$\chi_2^*(x,k)e^{-ikx/2} = 1 - c\int \frac{dp_1}{2\pi}\frac{dk_1}{2\pi}\frac{e^{i(k_1-p_1)x}}{(p_1-k-i\epsilon)(p_1-k_1-i\epsilon)}R^+(p_1)R(k_1) + \ldots \quad (3.28)$$

The final step in the Gel'fand-Levitan procedure is to recover the local field operator $\phi(x)$ by taking the $k\to\infty$ limit of the Jost solution,

$$\chi_1(x,k)e^{ikx/2} \xrightarrow[k\to\infty]{} -\frac{\sqrt{c}}{k}\phi(x) + O(1/k^2) \quad (3.29)$$

The field is thus written as an infinite series,

$$\phi(x) = \sum_{n=0}^{\infty} \phi^{(n)}(x) \quad (3.30)$$

where

$$\phi^{(n)}(x) = (-c)^n \int \left(\prod_1^n \frac{dp_i}{2\pi}\right)\left(\prod_0^n \frac{dk_i}{2\pi}\right) \frac{e^{i(\sum_0^n k_i - \sum_1^n p_i)x}}{\prod_{i=1}^n [(p_i-k_{i-1}-i\epsilon)(p_i-k_i-i\epsilon)]}$$

$$\times R^*(p_1)\ldots R^*(p_n)R(k_n)\ldots R(k_0) \quad (3.31)$$

The asymptotic expression for the other component of the Jost solution χ_2 yields a series for the charge density $j_0(x) = \phi^*(x)\phi(x)$.

Gel'fand-Levitan series as a generalized Jordan-Wigner transformation

The Gel'fand-Levitan transform (3.31) has a very interesting structure which can be studied term by term. Perhaps I should say at the outset that I'm not entirely satisfied with the style of analysis that I'll outline in this and subsequent sections. It would be nice if there were a more elegant way of studying Green's functions than term-by-term analysis of series expansions. My general feeling is that a better approach would make more direct use of the Gel'fand-Levitan integral equation and the Jost solutions, but such an approach has not yet been devised. The situation is reminiscent of the direct problem, where the properties of the a and b operators were first discovered by studying their series expansions and then subsequently derived by more elegant means. I hope that this history will repeat itself for the inverse problem, but for now I must rely on the term-by-term approach.

The lowest order term in (3.30) is just the Fourier transform of the reflection coefficient

$$\phi^{(0)}(x) = \int \frac{dk_0}{2\pi} e^{ik_0 x} R(k_0) \equiv \tilde{R}(x). \qquad (3.32)$$

The second term is

$$\phi^{(1)}(x) = \int \frac{dp_1}{2\pi}\frac{dk_0}{2\pi}\frac{dk_1}{2\pi} e^{i(k_0+k_1-p_1)x} \frac{(-c)R^*(p_1)R(k_1)R(k_0)}{(p_1-k_0)(p_1-k_1)}. \qquad (3.33)$$

Hereafter, momentum denominators will be understood to have infinitesimal negative imaginary parts. By writing the denominator in (3.33) as

$$\frac{1}{(p_1-k_0)(p_1-k_1)} = \frac{1}{k_{10}}\left[\frac{1}{p_1-k_0} - \frac{1}{p_1-k_1}\right], \qquad (3.34)$$

making the charge of variables $k_1 \leftrightarrow k_0$ in the first term and using the commutation relation (2.53) we can replace the integrand in (3.33) by

$$\frac{(-c)}{(p_1-k_0)(p_1-k_1)} + \frac{(-2c)}{(p_1-k_1)(k_{10}+ic)} = \frac{(-i)}{p_1-k_1}[S(k_{10}) - 1]. \qquad (3.35)$$

Equation (3.33) can then be written very simply in coordinate space

$$\phi^{(1)}(x) = \int_x^\infty dz\ [\tilde{R}^*(z)\tilde{R}(x)\tilde{R}(z) - \tilde{R}^*(z)\tilde{R}(z)\tilde{R}(x)]. \tag{3.36}$$

To understand the general term $\phi^{(n)}$ you should think of (3.36) as being obtained from $\phi^{(0)}(x)$ [i.e., $\tilde{R}(x)$] by inserting the operators $\tilde{R}^*(z)$ and $\tilde{R}(z)$ in two different ways and then integrating over z. The first term in (3.36) is an "outside" insertion and appears with a plus sign, while the second term is an "inside" insertion and has a minus sign. This pattern repeats itself in a straightforward way for the higher terms in the series, with each term $\phi^{(n)}$ being obtained from the previous term $\phi^{(n-1)}$ by an "outside minus inside" insertion of $\tilde{R}^*(z_n)$ and $\tilde{R}(z_n)$, with z_n integrated from z_{n-1} to ∞. For example, the next term is

$$\begin{aligned}\phi^2(x) = \int_x^\infty dz_1 \int_{z_1}^\infty dz_2 \{ & [\tilde{R}^*(z_2)\tilde{R}^*(z_1)\tilde{R}(x)\tilde{R}(z_1)\tilde{R}(z_2) \\ & - \tilde{R}^*(z_1)\tilde{R}^*(z_2)\tilde{R}(z_2)\tilde{R}(x)\tilde{R}(z_1)] \\ & - [\tilde{R}^*(z_2)\tilde{R}^*(z_1)\tilde{R}(z_1)\tilde{R}(x)\tilde{R}(z_2) - \tilde{R}^*(z_1)\tilde{R}^*(z_2)\tilde{R}(z_2)\tilde{R}(z_1)\tilde{R}(x)]\},\end{aligned} \tag{3.37}$$

where the first two terms in (3.37) are obtained from the first term in (3.36) and the second two terms of (3.37) are obtained from the second term in (3.36). The general term can be written most easily in momentum space,

$$\begin{aligned}\phi^{(n)}(x) = \int \prod_1^n \frac{dp_i}{2\pi} \prod_0^n \frac{dk_i}{2\pi}\, e^{ik_0x} \int dz_1 \dots dz_n\, \theta(x < z_1 < \dots < z_n) \\ \times \prod_{i=1}^n e^{i(k_i - p_i)z_i} (S_{10}-1)(S_{20}S^{12}S_{21}-1)\dots(S_{n0}S^{1n}S_{n1}\dots S^{n-1,n}S_{n,n-1}-1) \\ \times R^*(p_1)\dots R^*(p_n)R(k_n)\dots R(k_0),\end{aligned} \tag{3.38}$$

where $S^{ij} \equiv S(p_{ij})$ and $S_{ij} \equiv S(k_{ij})$. The factors (SS...S-1) in (3.38) are the momentum space version of "outside minus inside" insertions. I will not give a complete derivation of (3.38) here. The only derivation I know involves a rather lengthy combinatorial analysis which is most easily handled by graphical techniques. It turns out that the Gel'fand-Levitan series for $\phi(x)$ can be given a convenient graphical interpretation in terms of "factorized graphs," which were developed for this model several years ago.[19] This graphical formalism is very useful for handling the combinatorics involved in deriving formulas like (3.38), but it would take us too far afield to describe it here.

The form of the Gel'fand-Levitan series provided by Eq. (3.38) is particularly well suited to studying Green's functions in the strong coupling ($c\to\infty$) limit. In fact, in the limit $c\to\infty$, the Gel'fand-Levitan transform reduces to the more familiar Jordan-Wigner transformation.[7] For $c\to\infty$, $S\to-1$, and Eq. (3.38) reduces to

$$\phi^{(n)}(x) = \frac{(-2)^N}{N!} N_R \left[\int_x^\infty dz\ \tilde{R}^*(z)\tilde{R}(z)\right] \tilde{R}(x), \tag{3.39}$$

where N_R specifies normal ordering with respect to the R operators. Note that the algebra of R operators (2.53)-(2.54) reduces to canonical anti-commutation relations, and thus $\tilde{R}(x)$ is a local fermion field. The transform (3.30) reduces to

$$\phi(x) = N_R \exp\left[-2\int_x^\infty \tilde{R}^*(z)\tilde{R}(z)dz\right] \tilde{R}(x) \tag{3.40}$$

which may also be written

$$\phi(x) = \exp\left[i\pi\int_x^\infty \tilde{R}^*(z)\tilde{R}(z)dz\right] \tilde{R}(x). \tag{3.41}$$

This can be recognized as the standard form of a Jordan-Wigner fermion-to-boson transformation. The Jordan-Wigner transformation is also used in the solution of other models (e.g., the 2-D Ising model and the X-Y spin chain) which have the algebraic structure of a free fermion theory. These free fermion models can be regarded as special cases of more general Bethe's ansatz models (e.g., the Ising model and XY spin chain are special cases of the Baxter model and the XYZ spin chain respectively). Whereas the theory of Green's functions for Bethe's ansatz models is not very well understood, the special free-fermion cases are rather well-studied. The Green's functions for the $c=\infty$ non-linear Schrödinger model were first discussed by Schultz[20] and Lenard,[21] who related the 2n-point functions to the nth Fredholm minor associated with an integral kernel $K(x,y) = \sin(x-y)/(x-y)$. I'll come back to this result in Section V, where I shall discuss the large c expansion of the Green's functions. Recently, in an elegant series of developments by Sato, Miwa, and Jimbo,[22] the Green's functions for the free-fermion models were found to be deeply related to the theory of isomonodronic deformations of linear differential equations, whose mathematical origins go back to the early part of this century. In particular, this connection allowed SMJ to express the two-point functions for these models in closed form in terms of Painlevé functions. I'll review some of these developments in the last lecture.

IV. GREEN'S FUNCTIONS-GENERAL FORMALISM

I want to consider the two-point equal time correlation function $\langle \phi^*(x)\ \phi(y) \rangle_\Omega$ where $\langle \ldots \rangle_\Omega$ represents either a ground state expectation value or a thermal average, depending on whether we're discussing the zero or finite-temperature Green's functions. The basic idea of the quantum inverse approach to Green's functions is to express $\phi^*(x)$ and $\phi(y)$ in terms of R* and R operators and use this expression to compute expectation values. In order to do this we need two theorems, a reordering theorem and a trace theorem. The reordering theorem tells how to write the operator product $\phi^*(x)\ \phi(y)$ as a normal ordered functional of R* and R. The trace theorem tells how to compute the thermal average of a normal product of R*'s and R's.

Reordering Theorem

Beginning with the Gel'fand Levitan Series

$$\phi(y) = \sum_{N=0}^{\infty} \int \prod_{1}^{N} \frac{dp_i}{2\pi} \prod_{0}^{N} \frac{dk_i}{2\pi}\, g_N\, (p_i, k_i; y)$$

$$R^*(p_1) \ldots R^*(p_N)\ R(k_N) \ldots R(k_0) \tag{4.1}$$

$$\phi^*(x) = \sum_{N=0}^{\infty} \int \prod_{0}^{N} \frac{dp_i}{2\pi} \prod_{1}^{N} \frac{dk_i}{2\pi}\, g_N^*(k_i, p_i; x)$$

$$R^*(p_0) \ldots R^*(p_N)\ R(k_N) \ldots R(k_1), \tag{4.2}$$

we can form the operator product $\phi^*(x)\ \phi(y)$ which must be rearranged into normal ordered form. For the present discussion, the integrand g_N may be taken from either (3.31) or (3.38). Recall that in the derivation of the quantum Gel'fand-Levitan equation we made essential use of the fact that the commutator $[R^*(k), \phi^*(x)]$, Eq. (3.17), could be analytically continued into the lower half-plane. This same analyticity property can be used to derive a reordering theorem for $\phi^*(x)\ \phi(y)$. For definiteness, consider the case $x > y$. Write $\phi^*(x)$ as a GL series but leave $\phi(y)$, giving

$$\phi^*(x)\ \phi(y) = \sum_{N=0}^{\infty} \int \prod_{0}^{N} \frac{dp_i}{2\pi} \prod_{1}^{N} \frac{dk_i}{2\pi}\, g_N^*\,(k_i, p_i; x)$$

$$R^*(p_0)\ldots R^*(p_N) R(k_N)\ldots R(k_1)\,\phi(y). \tag{4.3}$$

For $x > y$, the analyticity of $[R(k), \phi(y)]$ in the upper half-plane allows us to move $\phi(y)$ to the left past all the $R(k)$'s. All the commutator terms vanish, since for each k_i $g_N^*(k_i,p_i)$ is also analytic in the upper half-plane and the integrand $\to 0$ asymptotically. Thus $\phi(y)$ can be placed between the R^*'s and the R's in (4.3) and then expanded, yielding a normal ordered series for $\phi^*(x)\ \phi(y)$:

$$\phi^*(x)\ \phi(y) = \sum_{N=0}^{\infty} \int \prod_{0}^{N} \frac{dp_i}{2\pi} \prod_{0}^{N} \frac{dk_i}{2\pi}\, F_N(p_i, k_i; x, y)$$

$$R^*(p_0)\ldots R^*(p_N) R(k_N)\ldots R(k_0), \tag{4.4}$$

where

$$F_N(p_i,k_i;x,y) = \sum_{\ell=0}^{N} g_\ell^*(k_0\ldots k_{\ell-1}; p_0\ldots p_\ell; x)\ g_{n-\ell}(p_{\ell+1}\ldots p_N; k_\ell\ldots k_N; y) \tag{4.5}$$

Temperature Green's Functions: Trace Theorem

As I discussed in Section II, the theory of Green's functions is being developed here in the infinite volume formalism, where the R operators have simple commutation relations (2.53)-(2.54) and also commute simply with the Hamiltonian

$$[H,R^*(k)] = k^2R^*(k). \tag{4.6}$$

With these algebraic properties, the two-point function can be computed term by term from the series (4.4). Of course, in the infinite volume formalism we must be careful to define the correct prescription for handling the infrared singularities which arise from integrations over an infinite volume. The subtleties associated with these infrared singularities and their relevance to the formulation of statistical mechanics without a box were discussed some time ago in the language of factorized graphs. (See the second paper in Ref. 19.) The procedure which I will outline below evolved from these graphical studies.

We will consider the finite temperature Green's function

$$G_{\beta,\mu}(x - y) = \frac{\mathrm{Tr}\left[\phi^*(x)\ \phi(y)e^{-\beta\Omega}\right]}{\mathrm{Tr}e^{-\beta\Omega}}, \tag{4.7}$$

where $\Omega = H-\mu N$. In computing the trace in the numerator of (4.7) we must consider diagonal matrix elements of the operator in square brackets. In the calculation of these matrix elements, we encounter two basic types of infrared divergence. One type arises from the presence of disconnected graphs in the trace, which leads to momentum space delta-functions with vanishing argument (i.e., $\delta(0)$ factors). These disconnected graphs may be easily removed by dividing out a factor of $\mathrm{Tr}e^{-\beta\Omega}$ as in (4.7). The other type of infrared divergence is more subtle. It arises in the connected part of a matrix element, e.g.,

$$\langle p_1 \cdots p_N | \phi^*(x)\ \phi(y)\ e^{-\beta\Omega} | k_1 \cdots k_N \rangle, \tag{4.8}$$

when we try to take the forward limit $p_i \to k_i$. Imagine computing the matrix element (4.8) from the many-body coordinate space wave functions. This involves integrating over the coordinates $z_1, \ldots, z_N$ of the N particles. The forward singularities arise from the asymptotic parts of this integration which become undamped in the limit $p_i \to k_i$ (i.e., terms which behave like $e^{i(p_i - k_i)z_i}$ as $z_i \to \pm\infty$). If we considered a limit of (4.8) where some but not all of the p_i's are set equal to the k_i's, then these singularities are really there and the limit does not exist because of divergences of the form $(p_i - k_i \pm i\varepsilon)^{-1}$. However, to compute a trace we need the diagonal (forward) matrix element, which is obtained by setting all momentum differences $(p_i - k_i)$ to zero simultaneously with fixed ratio. For the connected part of the matrix element (4.8) this forward limit is finite, because each singular denomintor $(p_i - k_i \pm i\varepsilon)$ is multiplied by a vanishing factor of the form $[e^{i(\Theta-\Theta')} - 1]$, where Θ is a sum of Bethe's ansatz phase shifts depending on the relative momenta k_{ij} in the initial state, and Θ' is the corresponding sum of phase shifts for the p_{ij}'s in the final state. In the forward limit the phase shifts in the initial and final states match up, i.e. $\Theta' \to \Theta$, rendering this limit finite. Pursuing this argument, it is now easy to see the correct procedure for calculating the forward matrix elements needed to compute the trace in (4.7). Consider the effect of any reasonable sort of cutoff on the coordinate space integrations, e.g., a sharp cutoff (a box of length L) or an adiabatic cutoff (i.e., keeping the $i\varepsilon$'s finite in the singular denominators). This simply regularizes the singular denominators without affecting the vanishing numerators $[e^{i(\theta-\theta')}-1]$. So the correct prescription is to set to zero all terms in the matrix element which have singular factors of the form

$$\frac{\left[e^{i(\theta-\theta')}-1\right]}{p_i-k_i \pm i\varepsilon}. \tag{4.9}$$

The connected forward matrix element is given by the remaining terms which have no singular ratios and hence have an unambiguous forward limit.

I will now introduce the basic device that will be used to correctly regularize the infrared singularities and compute the temperature Green's function (4.7). I will call this device the "infinitesimal boost method." Define the Galilean boost generator

$$K = \int x\phi^*(x)\ \phi(x)\ dx. \tag{4.10}$$

The R operators have a simple behavior under boosts:

$$e^{iqK}R(k)e^{-iqK} = R(k + q). \tag{4.11}$$

The basic assertion of the infinitesimal boost method is that the Green's function (4.7) is given by the formula

$$G_{\beta,\mu}(x-y) = \lim_{q\to 0} \mathrm{Tr}\left[\phi^*(x)\ \phi(y)\ e^{-\beta\Omega}e^{-iqK}\right]. \tag{4.12}$$

From the previous discussion it is easy to see why this method works. A forward N-body matrix element of the operator in square brackets in (4.12) will be of the form (4.8), where the k_i's are shifted from the p_i's by a small momentum q, i.e., $k_i = p_i - q$. This does two things. First is eliminates disconnected graphs, since a disconnected subgraph is essentially an integrated matrix element of $e^{-\beta\Omega}$ between states $\langle p_1,\ldots,p_\ell|$ and $|p_1-q,\ldots,p_\ell-q\rangle$ which vanishes by momentum conservation. (In fact we could have divided (4.12) by a factor $Tr(e^{-\beta\Omega}e^{-iqK})$, which is unity because it only receives a contribution from the zero particle state.) The fully connected graphs do not vanish because the operator $\phi^*(x)\ \phi(y)$ is there to absorb the momentum Nq. In addition to eliminating disconnected graphs, formula (4.12) also sets factors like (4.9) in the singular connected graphs to zero (which, as I have argued, is the correct thing to do). This happens because the phase shifts Θ and Θ' in the numerator depend only on the relative momenta k_{ij} and p_{ij} which are not affected by a Galilean boost. Thus the numerator is identically zero even for finite q.

Using the equation (4.12) along with the series (4.4) and the algebraic properties of the R-operators, we may compute the Green's function term by term in the series. To do this we will use a convenient theorem for evaluating traces of the form $Tr\{R^*(p_0)\ldots R^*(p_N)R(k_N)\ldots R(k_0)e^{-\beta\Omega}$ $e^{-iqK}\}$. Consider first the simplest case N=0,

$$Tr\{R^*(p)\ R(k)\ e^{-\beta\Omega}\ e^{-iqK}\}. \tag{4.13}$$

Using the properties

$$R(k)\ e^{-\beta\Omega} = e^{-\beta(k^2-\mu)}\ e^{-\beta\Omega}\ R(k) \tag{4.14}$$

$$R(k)\ e^{-iqK} = e^{-iqK}\ R(k + q) \tag{4.15}$$

$$R(k)\ R^*(p) = 2\pi\delta(p - k) + S(k - p)\ R^*(p)\ R(k), \tag{4.16}$$

and the cyclic property of the trace, we generate a fugacity series for (4.13) in the limit $q \to 0$:

$$\mathrm{Tr}\,\{R^*(p)R(k)\ e^{-\beta\Omega}\ e^{-iqK}\} \xrightarrow[q\to 0]{} -\sum_{n=1}^{\infty} (-z)^n\ e^{-n\beta k^2} \times \langle k-nq|p\rangle\ \{1+O(q)\}, \tag{4.17}$$

where $z \equiv e^{\beta\mu}$ = fugacity. By an inductive argument, this result can be generalized to the following trace theorem:

$$\mathrm{Tr}\,\{R^*(p_0)\ldots R^*(p_N)\ R(k_N)\ldots R(k_0)\ e^{-\beta\Omega}\ e^{-iqK}\}$$

$$= (-1)^N \sum_{n_0,n_1,\ldots,n_N=1}^{\infty}\ \prod_{i=0}^{N} \left[(-z)^{n_i}\ e^{-n_i\beta k_i^2}\right] \tag{4.18}$$

$$\langle k_0 + n_0 q,\ldots,k_N + n_N q|p_0\ldots p_N\rangle \times \{1 + O\ (q)\},$$

where

$$|p_0\ldots p_N\rangle \equiv R^*(p_0)\ldots R^*(p_N)|\ 0\rangle. \tag{4.19}$$

In normal ordered GL series, such as (4.1) or (4.4), the integrands g_N or F_N are not uniquely specified. This is clear, for example, from the equivalence of expressions (3.31) and (3.38). What is uniquely specified is the "R-symmetrized" function $g_N^{(S)}$ or $F_N^{(S)}$ which is obtained by symmetrizing over the p's and over the k's and using the commutation relations of the R-operators. Thus, for example

$$F_N^{(S)}(p,k) = \frac{1}{[(N+1)!]} \sum_{P,Q} F_N(Pp, Qk) \prod_{\substack{i<j \\ P_i>P_j}} S(p_{ij}) \prod_{\substack{i<j \\ Q_i>Q_j}} S(k_{ji}). \tag{4.20}$$

Any two functions which lead to the same R-symmetrized function will give equivalent operator expressions. Using the trace theorem (4.18) and the GL series (4.4), and writing the inner product of R-states in (4.18) as sums of products of δ-functions and S-matrices we get the result for the Green's function,

$$G_{\beta,\mu}(x-y) = \lim_{q\to 0} \sum_{N=0}^{\infty} \sum_{n_0,\ldots n_N=1}^{N} (-1)^N \int \prod_{i=0}^{N} \left[(-z)^{n_i} e^{-\beta n_i p_i^2} \frac{dp_i}{2\pi} \right] \times \tag{4.21}$$

$$\times F_N^{(S)}(p, p - nq;\ x,y).$$

In order to proceed further, we must derive some properties of the functions $F_N^{(S)}$. These are obtained from (4.5) and (4.20). Note that the integrands g_n in the GL series for $\phi(x)$ can be written in many different ways (only the R-symmetrized function has meaning), in particular, as in (3.31) or (3.38). Let us first consider the n_i

dependence of $F_N^{(S)}(p,p-nq;\ x,y)$. Since the denominators in (3.31) involve only single momentum differences (p_i-k_j), we conclude that

$$F_N^{(S)}(p,p-nq;\ x,y) \xrightarrow[q\to 0]{} \frac{H_N(n_0,\cdots n_N;\ p,x,y)}{n_0 n_1\cdots n_N}, \tag{4.22}$$

where H_N is a homogeneous $(N+1)^{th}$ order multinomial in $n_0,\cdots,n_N$ which is symmetric under simultaneous permutation of p_i's and n_i's. (The finiteness of $F_N^{(S)}$ in the $q \to 0$ limit follows from the inductive argument outlined below.) Thus each term in H_N is of the form

$$n_0^{\lambda_0} n_1^{\lambda_1} \cdots n_N^{\lambda_N} \times \text{function of } (p,x,y), \tag{4.23}$$

where

$$\sum_{i=0}^{N} \lambda_i = N + 1. \tag{4.24}$$

Let us pick out the "nonsingular" (i.e., n_i independent) term in (4.22) by writing

$$H_N(n_0\cdots n_N;\ p,x,y) = n_0 n_1\cdots n_N\ f_N(p;x,y) + \tilde{H}_N(n_0\cdots n_N;p,x,y), \tag{4.25}$$

where $\tilde{H}_N$ contains only terms where one or more of the λ_i's is zero. The point of making this separation is that now $\tilde{H}_N$ may be obtained by symmetry in the n_i's from its value with one of the n_i's set equal to zero, e.g., $\tilde{H}_N|_{n_N=0}$. The function $\tilde{H}_N|_{n_N=0}$ is determined by the residue of the pole in $F_N^{(S)}(p,k;x,y)$ at $p_N = k_N$ which can be related to the lower order function $F_{N-1}^{(S)}$. This is the essential inductive step which allows us to sum up all the n_i-dependent terms in (4.21) and express the Green's function entirely in terms of the functions f_N in (4.25). To study the residue of the pole in $F_N^{(S)}$ at $p_N = k_N$, it is convenient to use (4.20) with an unsymmetrized function F_N which is obtained from (3.38):

$$F_N(\mathbf{p},\mathbf{k};x,y) = e^{ik_0y-p_0x} \sum_{\ell=0}^{N} \int dz_1\ldots dz_N\ \theta(y<z_1<\ldots<z_\ell<x<z_{\ell+1}<\ldots<z_N)$$

$$\times \left[\prod_{i=1}^{N} e^{i(k_i-p_i)z_i}\right] (S_{10}-1)\ldots(S_{\ell 0}S^{1\ell}S_{\ell 1}\ldots S^{\ell-1,\ell}S_{\ell,\ell-1}-1) \qquad (4.26)$$

$$\times (S^{0,\ell+1}S_{\ell+1,0}\ldots S^{\ell,\ell+1}S_{\ell+1,\ell}-1)\ldots(S^{0N}S_{N0}\ldots S^{N-1,N}S_{N,N-1}-1).$$

From this expression it is easy to show that the residue at p_N-k_N is given by

$$F_N^{(S)} \underset{p_N \to k_N}{\longrightarrow} \frac{(-i)}{p_N - k_N} \left\{\prod_{i=0}^{N-1} S(p_i-p_N)\ S(p_N - k_i) -1\right\} F_{N-1}^{(S)}$$

$$\underset{k_i = p_i - n_i q}{\longrightarrow} \left\{(\frac{-i}{n_N}) \sum_{i=0}^{N-1} n_i \Delta(p_i-p_N)\right\} F_{N-1}^{(S)}. \qquad (4.27)$$

where $\Delta(k)$ is given in (2.18). The relation (4.27) allows us to sum up the n_i dependent terms in (4.21). To understand the result, it is instructive to first consider the result of summing only the nonsingular (n_i-independent) terms in $F_N^{(S)}$ (i.e., keeping only the first term in (4.25). This would give

$$\sum_{N=0}^{\infty} \prod_{i=0}^{N} \left[\frac{1}{e^{\beta(p_i^{\,2}-\mu)} + 1} \frac{dp_i}{2\pi} \right] f_N(p;x,y). \tag{4.28}$$

Using the induction (4.27), it can be shown that the sole effect of the singular terms is to replace $\left[e^{\beta(p_i^2-\mu)} + 1\right]^{-1}$ by $\left[e^{\beta\varepsilon(p_i)} + 1\right]^{-1}$ where $\varepsilon(p)$ is the excitation energy function of Yang and Yang, Eq. (2.21). Thus, the full Green's function is reduced to

$$G_{\beta,\mu}(x-y) = \sum_{N=0}^{\infty} \int \prod_{i=0}^{N} \left[\tilde{\rho}(p_i) \frac{dp_i}{2\pi}\right] f_N(p;x,y), \tag{4.29}$$

where

$$\tilde{\rho}(p) = \frac{1}{e^{\beta\varepsilon(p)} + 1}. \tag{4.30}$$

We have now isolated all the β and μ dependence of $G_{\beta,\mu}(x-y)$ in terms of a single particle distribution function $\tilde{\rho}(p)$. The familiar thermodynamics of Yang and Yang can be recovered from (4.29) in the limit $(x-y) \to 0$. In this case the polynomial H_N in (4.25) can be shown to have an overall factor of $\sum_{i=0}^{N} n_i$. This fact allows us to determine f_N as well as $\tilde{H}_N$ from the induction (4.27), giving

$$G_{\beta,\mu}(0) = \sum_{N=0}^{\infty} \int \prod_{i=0}^{N} \left[\tilde{\rho}(p_i) \frac{dp_i}{2\pi}\right] \prod_{i=0}^{N-1} \Delta(p_i - p_{i+1}). \tag{4.31}$$

This can be recognized as the expansion of the Yang and Yang integral equation for the density function $\rho(k)$

$$\frac{\rho(k)}{\tilde{\rho}(k)} = 1 + \int \frac{dq}{2\pi} \Delta(k - q) \rho(q). \tag{4.32}$$

Thus, the zero separation Green's function, which is just $\langle\phi^*(0)\phi(0)\rangle_\Omega$ = particle density, is given by

$$G_{\beta\mu}(0) = \int dk \; \rho(k), \tag{4.33}$$

where $\rho(k)$ is defined by (4.32), (4.30), and (2.21). From the density as a function of β and μ, other thermodynamic quantities may be derived.

Finally, let me note for later reference that the Green's function expression (4.29) has a simple zero temperature limit

$$G(x-y) = \lim_{\beta\to\infty} G_{\beta,\mu}(x-y) = \sum_{N=0}^{\infty} \int_{-k_F}^{k_F} \prod_{0}^{N} \frac{dp_i}{2\pi} f_N(p;x,y). \tag{4.34}$$

To summarize, we compute the Green's functions as follows: Begin with the functions $F_N^{(S)}$, the R-symmetrized GL integrands for $\phi^*(x)\phi(y)$ defined by (4.20) with some suitable unsymmetrized integrands F_N, e.g., Eq. (4.26). Then calculate the functions $f_N(p;x,y)$ by Eqs. (4.22) and (4.25). The zero-temperature and finite temperature Green's functions are given by (4.34) and (4.29) respectively.

It would be nice to write the Green's function in closed form, but so far this has not been done. But recently it was shown that the first two terms in a strong coupling (large c) expansion of G(x-y) may be expressed in closed form in terms of Painlevé functions. The large coupling results and their connection with the work of Sato, Miwa, and Jimbo will be discussed in the last two lectures.

V. LARGE c EXPANSION OF THE TWO-POINT FUNCTION

The Green's functions for impenetrable bosons ($c=\infty$) were extensively studied, first by Schultz and Lenard and more recently by Jimbo et al. To make contact with these results we will consider a large c expansion of the two-point function

$$G = G^{(0)} + G^{(1)} + G^{(2)} + \dots, \tag{5.1}$$

where $G^{(n)}$ is of order $(1/c)^n$. The form of the GL integrands F_N given in (4.26) is well-suited for studying the $c\to\infty$ limit. Note that the z_i integrations are ordered, $z_1 < z_2 < \dots < z_N$, and that ℓ of the integrations are "trapped" between y and x, and $N-\ell$ of them are "untrapped" between x and ∞. With each trapped z-integration is associated a factor

$$(\text{odd number of } S\text{'s} - 1), \tag{5.2}$$

while each untrapped z-integration has a factor

$$(\text{even number of } S\text{'s} - 1). \tag{5.3}$$

In the limit $c\to\infty$, $S(p_{ij}) \to -1$. and the factors (5.2) and (5.3) become -2 and 0 respectively. Thus, only the terms in (4.26) with all z-integrations trapped contribute at $c = \infty$. (The Jordan-Wigner "tails" in

(3.41) cancel exactly outside the interval $y < z < x$.) This gives

$$F_N(p,k;x,y) = (-2)^N \int dz_1 \ldots dz_N \theta(y < z_1 < \ldots < z_N < x) \prod_{i=1}^{N} e^{i(k_i - p_i)z_i}. \quad (5.4)$$

Since the R-operators anticommute for $c = \infty$, we may symmetrize (5.4) over simultaneous permutations of p_i's and k_i's, allowing us to make the replacement

$$F_N(p,k,x,y) \to \frac{(-2)^N}{N!} \int_y^x dz_1 \ldots \int_y^x dz_N \prod_{i=1}^{N} e^{i(k_i - p_i)z_i}. \quad (5.5)$$

Note that because there are no untrapped integrations, there are no poles at $p_i = k_i$ and we may set $q = 0$ from the start. Since $S = -1$, the R-symmetrized function $F_N^{(S)}$ involves a determinant. In this way we get the expression for the $c = \infty$ Green's function

$$G^{(0)}_{\beta,\mu}(x-y) = \sum_{N=0}^{\infty} \frac{(-2)^N}{N!} \int \left[\prod_0^N \tilde{\rho}(p_i) \frac{dp_i}{2\pi}\right] \int_y^x dz_1 \ldots \int_y^x dz_N \mathcal{D}_N(x,y;z,p), \quad (5.6)$$

where $\mathcal{D}_N$ is an $(N+1) \times (N+1)$ determinant of exponentials, e.g,

$$\mathcal{D}_0 = 1 \quad (5.7a)$$

$$\mathcal{D}_1 = \begin{vmatrix} e^{-ip_0(x-y)} & e^{-ip_0(x-z_1)} \\ e^{-ip_1(z_1-y)} & 1 \end{vmatrix} \quad (5.7b)$$

$$\mathcal{D}_2 = \begin{vmatrix} e^{-ip_0(x-y)} & e^{-ip_0(x-z_1)} & e^{-ip_1(x-z_2)} \\ e^{-ip_1(z_1-y)} & 1 & e^{-ip_1(z_1-z_2)} \\ e^{-ip_2(z_2-y)} & e^{-ip_2(z_2-z_1)} & 1 \end{vmatrix} \quad (5.7c)$$

etc.

By carrying out the p_i integrations in (5.6) we may write

$$G^{(0)}_{\beta,\mu}(x-y) = \frac{1}{2}\left\{\lambda K(x,y) - \int_y^x dz_1 \begin{vmatrix} K(x,y) & K(x,z_1) \\ K(z,y) & K(z_1,z_1) \end{vmatrix} + \frac{\lambda^3}{2!}\int_y^x dz_1 \int_y^x dz_2 \, |3\times 3| - \ldots \right. \tag{5.8}$$

where $\lambda = 2/\pi$, and the kernel K is the Fourier transform of a Fermi-Dirac distribution,

$$K(x,y) = \frac{1}{2}\int e^{-ip(x-y)} \tilde{\rho}_0(p)\, dp \tag{5.9}$$

with $\rho_0(p) = [e^{\beta(p^2-\mu)} + 1]^{-1}$. At zero temperature the kernel reduces to

$$K(x-y) = \frac{1}{2}\int_{-k_F}^{k_F} e^{-ip(x-y)} dp = \frac{\sin(x-y)}{x-y}. \tag{5.10}$$

(Hereafter, I will set $k_F = 1$.)

The $c = \infty$ Green's function (5.8) is essentially a Fredholm minor associated with the integral kernel K(x,y). Let me remind you how an integral equation is solved by Fredholm determinants. Consider the integral equation for a function R(x,y),

$$R(x,y) = \lambda K(x,y) + \lambda\int_a^b dz\, K(x,z)\, R(z,y). \tag{5.11}$$

By a continuum version of Kramer's rule, R(x,y) may be written as a ratio of determinants,

$$R(x,y) = \frac{D_1(x,y;a,b)}{D(a,b)}, \tag{5.12}$$

where D_1 is the Fredholm minor

$$D_1(x;y;a,b) = \lambda K(x,y)-\lambda^2\int_a^b dz_1 \begin{vmatrix} K(x,y) & K(x,z_1) \\ K(z_1,y) & K(z_1,z_1) \end{vmatrix} + \frac{\lambda^3}{2!}\int_a^b dz_1 \int_a^b dz_2 |3\times 3| - \dots \tag{5.13}$$

and D is the Fredholm determinant,

$$D(a,b) = 1-\lambda\int_a^b K(z,z)dz + \frac{\lambda^2}{2!}\int_a^b dz_1 \int_a^b dz_2 \begin{vmatrix} K(z_1,z_1) & K(z_1,z_2) \\ K(z_2,z_1) & K(z_2,z_2) \end{vmatrix} - \dots \tag{5.14}$$

$$= \mathrm{Det}\ (1 - \lambda K).$$

The $c = \infty$ Green's function (5.8) is thus a Fredholm minor with its arguments evaluated at the endpoints of the integration region,

$$G^{(0)}(a - b) = \frac{1}{2}\, D_1(a,b;a,b). \tag{5.15}$$

All of these $c = \infty$ results have been known since the work of Lenard. Here we see how they follow as a special case of the quantum inverse formalism. Moreover, we can now go on to consider finite c corrections.

An expansion in powers of (1/c) can be obtained by collecting the terms in (4.26) according to the number of untrapped z-integrations. Each untrapped integration is accompanied by a factor (even number of S's - 1), which is of order (1/c). Here I'll consider the Green's function up to order (1/c), so only terms with zero or one untrapped integration must be kept.

$$G^{(1)} = G_0^{(1)} + G_1^{(1)}. \tag{5.16}$$

For the terms with no untrapped integrations, we use expansions of the form

$$(S_{10}-1) \sim (-2)(1-\frac{k_{10}}{ic})$$

$$(S_{10}-1)(S_{20}S^{12}S_{21}-1) \sim (-2)^2 \left[1-\frac{(k_{10}+k_{20}+k_{21}+p_{12})}{ic}\right] \quad (5.17)$$

etc. By this approach we obtain the 1/c correction

$$G_0^{(1)} = \left(\frac{1}{ic}\right) \sum_{N=1}^{\infty} \frac{(-2)^N}{N!} \int \left[\prod_{i=0}^{N} \tilde{\rho}(p_i) \frac{dp_i}{2\pi} \sum_{j=1}^{N} (p_0-p_j)\right] \times \int_y^x dz_1 \cdots \int_y^x dz_N \mathscr{D}_N(x,y;z,p), \quad (5.18)$$

and $\mathscr{D}_N$ is the same determinant of exponentials which appeared in the c = ∞ Green's function. By writing out the determinant in (5.18) and expressing the factor $\Sigma(p_0-p_j)$ as derivatives with respect to x,y, and z_i, it is possible to write (5.18) in terms of the Fredholm resolvent and minor (5.12) and (5.13). Let us define the functions R(t) and $D_1(t)$ by

$$R(t) = \lambda K(t,0) + \lambda^2 \int_0^t dz_1\, K(t,z_1)\, K(z_1,0) + \lambda^3 \int_0^t dz_1 \int_0^t dz_2\, K(t,z_1)\, K(z_1,z_2)\, K(z_2,0) + \cdots \quad (5.19)$$

$$D_1(t) = \lambda\, K(t,0) - \lambda^2 \int_0^t dz_1 \begin{vmatrix} K(t,0) & K(t,z_1) \\ K(z_1,0) & K(z_1,z_1) \end{vmatrix} + \frac{\lambda^3}{2!} \int_0^t dz_1 \int_0^t dz_2 |3 \times 3| - \cdots$$

(5.20)

After some manipulation, (5.18) may be written

$$G_0^{(1)}(t) = \frac{1}{\pi c} D_1 \left\{ \left(\frac{\partial^2 \ln R}{\partial t \partial \lambda} - \frac{\partial \ln R}{\partial t} \frac{\partial \ln R}{\partial \lambda} \right) - \left(\frac{\partial^2 \ln D_1}{\partial t \partial \lambda} - \frac{\partial \ln D_1}{\partial t} \frac{\partial \ln D_1}{\partial \lambda} \right) \right\} . \quad (5.21)$$

Next we must consider the contribution of the terms in (4.26) with one untrapped z-integration. The corresponding S-matrix factor may be expanded,

$$\left(S^{ON}S_{NO}\cdots S^{N-1,N}S_{N,N-1}\right) \sim \frac{2}{ic}\sum_{i=0}^{N-1} [(k_i - k_N) - (p_i - p_N)]$$
$$= \frac{2}{ic}\sum_{i=0}^{N} (k_i - p_i) + \frac{2N}{ic}(p_N - k_N). \tag{5.22}$$

Since this is already of order 1/c, the rest of the expression can be evaluated at $c = \infty$. In particular, the R's can be taken to anti-commute. The second term in (5.22) is found to vanish by antisymmetry, while the first term gives a contribution proportional to the $c = \infty$ Green's function,

$$G_1^{(1)} = \frac{2}{\pi c} G^{(0)}. \tag{5.23}$$

To summarize, the first two terms in a 1/c expansion of G(t) are

$$G^{(0)} = \frac{1}{2} D_1 \tag{5.24}$$

$$G^{(1)} = \frac{2}{\pi c} G^{(0)}\left\{1 + \left(\frac{\partial^2 \ln R}{\partial t \partial \lambda} - \frac{\partial \ln R}{\partial t}\frac{\partial \ln R}{\partial \lambda}\right) - \left(\frac{\partial^2 \ln D_1}{\partial t \partial \lambda} - \frac{\partial \ln D_1}{\partial t}\frac{\partial \ln D_1}{\partial \lambda}\right)\right\}, \tag{5.25}$$

where $R(t)$ and $D_1(t)$ are given by (5.19) and (5.20). In the last lecture, I'll discuss the method for treating Green's functions developed by Sato, Miwa, and Jimbo[22] and applied to the $c = \infty$ nonlinear Schrödinger model by Jimbo et al.[12] This method gives closed forms for $R(t)$ and $D_1(t)$ in terms of Painlevé transcendents.

VI. GREEN'S FUNCTIONS AS PAINLEVÉ FUNCTIONS

Simple Derivation Of Painlevé Equation

Consider the resolvent $R(x,y)$ at zero temperature defined by (5.11),

$$R(x,y) = \lambda K(x,y) + \lambda^2 \int_a^b K(x,z)\, K(z,y)\, dz + \dots = \left[\frac{\lambda K}{1-\lambda K}\right](x,y) \quad ,(6.1)$$

where the kernel $K(x,y) = \sin(x-y)/(x-y)$. The x and y dependence of (6.1) may be written in a factorized form by defining

$$R_\pm(x) = e^{\pm ix} + \lambda \int_a^b dz_1 K(x,z_1) e^{\pm iz_1} + \lambda^2 \int_a^b dz_1 \int_a^b dz_2 K(x,z_1) K(z_1,z_2) e^{\pm iz_2}$$

$$+ \dots = \left[\frac{1}{1-\lambda K} E_\pm\right](x), \qquad (6.2)$$

where

$$E_\pm(x) = e^{\pm ix}. \qquad (6.3)$$

The quantity $R_+(x)\, R_-(y) - R_-(x)\, R_+(y)$ can be worked out term by term using

$$E_+(z)\, E_-(z') - E_-(z)\, E_+(z') = 2i(z-z')\, K(z,z'), \qquad (6.4)$$

and

$$(x-z_1) + (z_1-z_2) + \dots + (z_m-y) = (x-y). \qquad (6.5)$$

This gives the factorized expression

$$R(x,y) = \frac{\lambda\left[R_+(x)\ R_-(y) - R_-(x)\ R_+(y)\right]}{2i(x - y)}. \tag{6.6}$$

The series (6.2) may be differentiated term by term to obtain $\partial R_\pm/\partial x$. Since K is a difference kernel, we may replace $\frac{\partial}{\partial x} \to -\frac{\partial}{\partial z_1}$, integrate by parts, then replace $\partial/\partial z_1 \to -\partial/\partial z_2$, integrate by parts again, etc., until the derivitive is acting on $e^{\pm i z_N}$. The surface terms can also be summed, and we find

$$\frac{\partial R_\pm(x)}{\partial x} = \pm i\ R_\pm(x) + R(x,a)\ R_\pm(a) - R(x,b)\ R_\pm(b). \tag{6.7}$$

Using the factorization property (6.6), it is seen that the column vector

$$y(x) = \begin{pmatrix} R_+(x) \\ R_-(x) \end{pmatrix}, \tag{6.8}$$

satisfies a first order equation

$$\frac{\partial y(x)}{\partial x} = \left[\frac{A(a)}{x-a} - \frac{A(b)}{x-b} + C\right] y(x), \tag{6.9}$$

where

$$A(a) = \lambda \begin{pmatrix} R_+(a)\ R_-(a) & -R_+^2(a) \\ R_-^2(a) & -R_+(a)\ R_-(a) \end{pmatrix}, \tag{6.10}$$

and

$$C = \begin{pmatrix} i & 0 \\ 0 & -i \end{pmatrix}. \tag{6.11}$$

We may also derive an equation by differentiating $R_{\pm}$ with respect to a or b. The derivative acting on the upper or lower limit of integration gives terms which sum up in the same way as the surface terms in (6.7). In this way we get

$$\frac{\partial y}{\partial b} = \frac{A(b)}{x - b} y, \tag{6.12}$$

$$\frac{\partial y}{\partial a} = \frac{-A(a)}{x - a} y, \tag{6.13}$$

Eqs. (6.9), (6.12), and (6.13) can be written as a total differential relation,

$$dy = \Omega y, \tag{6.14}$$

where Ω is a differential form given by

$$\Omega = A(a)\, d\ell n(x-a) - A(b)\, d\ell n(x-b) + C dx. \tag{6.15}$$

The linear system of equations (6.14) is the fundamental property of the series (6.1) with kernel sin(x-y)/(x-y) which was discovered by Jimbo, et al. Later on, I will discuss how (6.14) follows directly from an isomonodromy property. But first let me derive the Painlevé expressions for R(t) and D(t), using (6.14). Let

$$\begin{aligned} x = b &= t/2 \\ a &= -t/2, \end{aligned} \tag{6.16}$$

Then $$d\ell n(x-a) = d\ell nt = dt/t$$

$$d\ell n(x-b) = 0 \tag{6.17}$$

$$dx = 1/2\ dt.$$

Denote by $r_\pm(t)$ the quantities (6.2) evaluated at (6.16). Then (6.14) can be written

$$r'_+ = \frac{i}{2} r_+ + \frac{\lambda}{2it} (r_+^2 - r_-^2) r_- \tag{6.18a}$$

$$r'_- = -\frac{i}{2} r_- + \frac{\lambda}{2it} (r_+^2 - r_-^2)\ r_+. \tag{6.18b}$$

Now introduce two functions $r(t)$ and $\psi(t)$ by

$$r_+ = e^{i\pi/4}\ r \cosh \frac{1}{2} \left(\psi - \frac{i\pi}{2}\right) \tag{6.19a}$$

$$r_- = e^{i\pi/4}\ r \sinh \frac{1}{2} \left(\psi - \frac{i\pi}{2}\right). \tag{6.19b}$$

Then Eqs. (6.18) reduce to

$$r^2 = \frac{t}{\lambda} (\psi' + \cosh \psi), \tag{6.20}$$

and

$$\psi'' + \frac{1}{t} \psi' + \frac{1}{t} \cosh \psi - \frac{1}{2} \sinh 2\psi = 0. \tag{6.21}$$

Eq. (6.21) is equivalent to a Painlevé equation of the fifth kind. It is convenient to make another change of variables

$$\sinh \psi = \cot \phi. \tag{6.22}$$

Then Eq. (6.21) becomes

$$\phi'' = \left[(\phi')^2 - 1\right] \cot \phi + \frac{(1 - \phi')}{t}. \tag{6.23}$$

The function $\phi(t)$ is completely specified by (6.23) along with the boundary condition

$$\phi(t) \sim t - \lambda t^2 + O(t^3) \text{ as } t \to 0, \tag{6.24}$$

which follows from the series expression (6.2).

All the functions needed to express the $c = \infty$ Green's function and the $1/c$ correction can be expressed in terms of $\phi(t)$. The function $R(t)$ defined by (5.19) is

$$R(t) = \frac{1 - \phi'}{2 \sin\phi}, \tag{6.25}$$

while $D_1(t)$, Eq. (5.20) satisfies

$$\frac{\partial \ln D_1}{\partial t} = \frac{t\left[(\phi')^2 - 1\right]}{4 \sin^2\phi} + \cot \phi - \frac{1}{t}. \tag{6.26}$$

Eq. (6.26) along with the condition

$$D_1(0) = \lambda, \tag{6.27}$$

specifices $D_1(t)$ completely. By numerical integration of Eq. (6.23), the functions $G^{(0)}(t)$ and $G^{(1)}(t)$, Eqs. (5.24) and (5.25), may be easily plotted. The long distance behavior of the Green's function may be studied using the asymptotic expansion of $\phi(t)$. The behavior of $\phi(t)$ as $t \to \infty$ depends critically on the value of λ. For $-\infty < \lambda < 1/\pi$, the asymptotic behavior is $\phi(t) \sim t + O(\ell nt)$ while for $1/\pi < \lambda < \infty$, it is $\phi(t) \sim -t + O(\ell nt)$. At the critical value $\lambda = 1/\pi$, $\phi(t)$ goes to a constant, $\phi(t) \sim \pi/2 + O(1/t)$. The $c = \infty$ Green's function $G^{(0)}$ is given in terms of $\phi(t)$ at $\lambda = 2/\pi$, while the $1/c$ correction $G^{(1)}$ involves $\phi(t)$ and its first λ-derivative at that point. It is amusing to note that the critical value $\lambda = 1/\pi$ also arises in a physical problem, that of determining the eigenvalue distribution of random matrices.

To study the long distance behavior of the two-point function, we use the asymptotic expansion

$$\phi(t) \sim -t + t_0 + k\ell nt + O(1/t), \tag{6.28}$$

where k and t_0 are λ-dependent constants. For a detailed discussion of the asymptotic analysis, I refer you to the literature.[9] Here, I will simply mention that the dominant effect of the $1/c$ correction is to alter the power-law falloff of the Green's function. The full Green's function $G(t)$ behaves like

$$G(t) \sim \text{const} \times t^{-\nu} [1 + O(1/t)], \tag{6.29}$$

where ν is a c-dependent constant. At $c = \infty$, Vaidya and Tracy showed that $\nu = 1/2$. Our result for the $1/c$ correction gives

$$\nu = \frac{1}{2} - \frac{2k_F}{\pi c} + O(\frac{1}{c^2}). \tag{6.30}$$

This agrees with a recent result of Haldane[23] and also of Popov, who obtained the value of ν for arbitrary c,

$$\nu = \frac{1}{2} [\rho(k_F)]^{-2}, \tag{6.31}$$

where $\rho(k_F)$ is the Lieb-Liniger density function at the Fermi surface.

Monodromy and Isomonodromic Deformation Theory[22]

The connection between Green's functions and Painlevé functions is particularly fascinating because of the elegant mathematical structure which can be associated with the Painlevé equations. This mathematical structure forms the basis of the analysis of Sato, Miwa, and Jimbo. Before introducing these ideas, let me explain what Painlevé did to get his name attached to these functions. In 1902 Painlevé[25] studied and solved the problem of classifying all second order ordinary differential equations of the form

$$y'' = f(y,y',t) \tag{6.32}$$

where f is an algebraic function, and with the requirement that the solutions should have no movable singularities. A movable singularity is one whose position depends on integration constants (i.e., on boundary conditions) and not just on the parameters in the equation. For example, the equation $y' = 1/2y$ has a movable singularity because its solution is $y = (t-a)^{1/2}$ where a is an arbitrary constant which doesn't appear in the differential equation. Painlevé showed that there were six kinds of equations of this form whose solutions could not be expressed in terms of elementary functions. These equations are known as Painlevé I-VI. Somewhat later Schlesinger[26] and Garnier[27] showed that all six Painlevé equations were obtained in a natural way as integrability conditions in the deformation theory of ordinary differential equations. The relevance of Painlevé equations to the theory of Green's functions was first exhibited by Wu et al.,[28] who showed that the spin-spin correlation function of the two-dimensional Ising model in the scaling limit could be expressed in terms of a solution to Painlevé III. Motivated by this result, Sato, Miwa, and Jimbo discovered a very elegant derivation of the correlation function which exploited the monodromy property of a certain expectation value of order and disorder operators. The result of Jimbo et al.,[12] for the $c = \infty$ nonlinear Schrödinger model was obtained by a similar technique.

To introduce the idea of monodromy and isomonodromic deformations, let us consider a linear problem of the form

$$\frac{dY}{dx} = \sum_{\nu=1}^{N} \frac{A_\nu}{x-a_\nu} Y, \tag{6.33}$$

where A_ν, $\nu = 1,\ldots,N$, are x independent $M \times M$ matrices, and $Y(x)$ is an $M \times M$ matrix solution satisfying some specified boundary condition, e.g., $Y(x_0) = I$. $Y(x)$ is not generally single-valued as we continue around the singularities at $a_1,a_2,\ldots a_N$. In general, traversing a closed curve around the singularity at $x = a_\nu$ will produce a linear transformation

$$Y(x) \longrightarrow Y(x)\, M_\nu. \tag{6.34}$$

where M_ν is an x-independent matrix called the monodromy matrix. Near $x = a_\nu$ we can write

$$Y(x) = \hat{Y}(x)\ (x-a_\nu)^{L_\nu}, \tag{6.35}$$

where L_ν is a constant matrix and $\hat{Y}(x)$ is nonsingular at $x = a_\nu$. The monodromy matrices are related to the L_ν's by

$$M_\nu = e^{2i\pi L_\nu}. \tag{6.36}$$

Looking at the $x = a_\nu$ pole of $\frac{\partial Y}{\partial x} Y^{-1}$ we see that

$$A_\nu = \hat{Y}(a_\nu)\ L_\nu\ \hat{Y}(a_\nu)^{-1}. \tag{6.37}$$

The problem of constructing the monodromy matrices from the differential equation (6.33) is somewhat analogous to the direct problem of scattering theory. The analog of the inverse problem, i.e., reconstructing the function Y(x) and the differential equation (6.33) from the monodromy "data," is known as the Riemann-Hilbert problem. Schlesinger[26] addressed this question by studying the behavior of the equation (6.33) under a variation of the positions of the singularities a_ν. Specifically, he allowed the coefficient matrices A_ν to depend on the a_ν's and asked what conditions would lead to monodromy matrices M_ν which were independent of the variation. Such an "isomonodromic deformation" leads to a linear system of equations for Y:

$$\frac{\partial Y}{\partial x} = \sum_{\nu=1}^{N} \frac{A_\nu}{x-a_\nu} Y \tag{6.38a}$$

$$\frac{\partial Y}{\partial a_\nu} = - \frac{A_\nu}{x-a_\nu} Y. \qquad (6.38b)$$

In differential forms, this reads

$$dY = \Omega Y \qquad (6.39)$$

where

$$\Omega = \sum_{\nu=1}^{N} A_\nu \, d\ell n(x-a_\nu). \qquad (6.40)$$

The dependence of the A_ν's on the a_μ's is given by the nonlinear consistency (integrability) conditions for the linear system (6.39). Using Poincare's Lemma, d^2(anything) = 0 (i.e., mixed partial derivatives taken in reverse order are equal), we get the integrability condition

$$d\Omega = \Omega\Lambda\Omega. \qquad (6.41)$$

This is analogous to $F_{\mu\nu} = 0$, Eq. (2.24) in the inverse scattering method. In explicit form, the dependence of the A_ν's on the a_μ's which yields fixed monodromy data is given by "Schlesinger's equations,"

$$\frac{\partial A_\nu}{\partial a_\mu} = \frac{\left[A_\nu, A_\mu\right]}{a_\nu - a_\mu} \quad (\mu \neq \nu) \qquad (6.42a)$$

$$\frac{\partial A_\nu}{\partial a_\nu} = - \sum_{\nu' \neq \nu} \frac{\left[A_\nu, A_{\nu'}\right]}{a_\nu - a_{\nu'}}. \qquad (6.42b)$$

In the simplest nontrivial case of Schlesinger's equations, the A_ν's are 2×2 matrices and there are N=4 singularities. This case reduces to an ordinary nonlinear differential equation which is just Painlevé VI. Garnier[27] showed that Painlevé I thru V could also be obtained as monodromy preserving deformation equations. For this one must consider the linear equation

$$\frac{\partial Y}{\partial x} = \left\{\sum_{\nu=1}^{N} \frac{A_\nu}{x-a_\nu} + C\right\} Y, \tag{6.43}$$

which allows for exponential behavior at infinity,

$$Y(x) \sim e^{Cx}\,[1 + O(1/x)]. \tag{6.44}$$

For (6.43) the simplest nontrivial case is N=2, and the deformation equations reduce to Painlevé V. This is the case which is relevant to the nonlinear Schrödinger model. The result expressing the functions $R(t)$ and $D_1(t)$, Eqs. (5.19) and (5.20) in terms of Painlevé function can be derived from a monodromy argument. To see how this works, separate the kernel $K(x,y) = \sin(x-y)/(x-y)$ into two pieces, $K = K_+ + K_-$, with

$$K_\pm(x,y) = \pm \frac{e^{\pm i(x-y)}}{2i(x-y)}. \tag{6.45}$$

Now define $R_\pm(x)$ as in (6.2), and also define the series

$$\bar{R}_\pm = E_\pm + \lambda K_- E_\pm + \lambda^2 K_- K E_\pm + \lambda^3 K_- K^2 E_\pm + \ldots = E_\pm + \lambda K_-\left[\frac{1}{1-\lambda K}\right] E_\pm. \tag{6.46}$$

We see that $R_\pm(x)$ is nonsingular at x=a and x=b, while $\bar{R}_\pm(x)$ has a cut in the x-plane from a to b. Using the discontinuity of the kernel,

$$\text{Disc } K_-(x,y) = \pi\delta(x-y), \tag{6.47}$$

we get

$$\text{Disc } \bar{R}_\pm(x) = \pi\lambda R_\pm(x). \tag{6.48}$$

Thus, the matrix

$$Y(x) = \begin{pmatrix} R_+(x) & \bar{R}_+(x) \\ R_-(x) & \bar{R}_-(x) \end{pmatrix}, \tag{6.49}$$

has an isomonodromy property. Defining closed curves γ_1 and γ_2 around a and b respectively, it is easy to show that

$$Y(x) \xrightarrow[\gamma_1]{} Y(x)M_1 \tag{6.50a}$$

$$Y(x) \xrightarrow[\gamma_2]{} Y(x)M_2, \tag{6.50b}$$

where the monodromy matrices are

$$M_1 = \begin{pmatrix} 1 & \pi\lambda \\ 0 & 1 \end{pmatrix} \qquad M_2 = \begin{pmatrix} 1 & -\pi\lambda \\ 0 & 1 \end{pmatrix}. \tag{6.51}$$

The monodromy is independent of the positions of the singularities a and b, and hence, by Schesinger's result, it follows that Y(x) obeys the linear relation

$$dY = \Omega Y \tag{6.52}$$

where

$$\Omega = A(a)\, d\ell n(x-a) - A(b)\, d\ell n(x-b) + C. \tag{6.53}$$

The first column of (6.52) is just Eq. (6.9), but now elegantly derived from monodromy properties. The Painlevé V equation follows from the integrability condition $d\Omega = \Omega\Lambda\Omega$.

The SMJ analysis of Green's functions adds some substance to the connection between integrability and duality. This is especially clear in the case of the two-dimensional Ising model, where the function which exhibits a monodromy property is constructed from expectation values of order and disorder fields, and the monodromy is a direct consequence of the algebraic properties of these fields. It is encouraging to note that very similar ideas have emerged in recent studies of four-dimensional gauge fields.[29]

Acknowledgments

The work described in these lectures was done in collaboration with Dennis Creamer and David Wilkinson, to whom I am grateful for many valuable conservations and insights.

REFERENCES

1. For reviews and more complete references, see Refs. 10 and 11.
2. E. K. Sklyanin and L. D. Faddeev, Dokl. Akad. Nauk SSSR **243**, 1430 (1978) [Sov. Phys.-Dokl. **23**, 902].
3. E. K. Sklyanin, Izv. Akad. Nauk SSSR **244**, 1337 (1979) [Sov. Phys.-Dokl. **24**, 107 (1979)].
4. H. B. Thacker and D. Wilkinson, Phys. Rev. **D19**, 3660 (1979).
5. J. Honerkamp, P. Weber, and A. Wiesler, Nucl. Phys. **B152**, 266 (1979).
6. D. B. Creamer, H. B. Thacker, and D. Wilkinson, Phys. Rev. **D21**, 1523 (1980).
7. D. B. Creamer, H. B. Thacker, and D. Wilkinson, Phys. Lett. **B92**, 144 (1980).
8. D. B. Creamer, H. B. Thacker, and D. Wilkinson, J. Math. Phys. **22**, 1084 (1981).
9. D. B. Creamer, H. B. Thacker, and D. Wilkinson, Phys. Rev. **D23**, and manuscript in preparation.
10. L. D. Faddeev, preprint LOMI P-2-79, to be published in English in Soviet Scientific Reviews, Section C (Mathematical Physics), Vol. 1, ed. S. P. Novikov (1981).
11. H. B. Thacker, Rev. Mod. Phys. **53**, 253 (1981).
12. M. Jimbo, T. Miwa, Y. Mori, and M. Sato, Physica **1D**, 80 (1980).
13. E. Lieb and W. Liniger, Phys. Rev. **130**, 1605 (1963); E. Lieb, Phys. Rev. **130**, 1616 (1963).
14. C. N. Yang and C. P. Yang, J. Math. Phys. **10**, 1115 (1969).
15. C. S. Gardner, J. M. Greene, M. D. Kruskal, and R. M. Miura, Phys. Rev. Lett. **19**, 1095 (1967).

16. V. E. Zakharov and A. B. Shabat, Zh. Eksp. Teor. Fiz. **61**, 118 (1971)[Sov. Phys.-JETP **34**, 62 (1972)].

17. R. J. Baxter, Ann. Phys. (N.Y.) **70**, 193 (1972).

18. L. A. Takhtajan and L. D. Faddeev, Usp. Mat. Nauk **34**, 13 (1979) [Russian Math. Survey **34**, 11 (1979)].

19. H. B. Thacker, Phys. Rev. **D14**, 3508 (1976); **D16**, 2515 (1977).

20. T. Schultz, J. Math. Phys. **4**, 666 (1963).

21. A. Lenard, J. Math. Phys. **5**, 930 (1964); **7**, 930 (1966).

22. M. Sato, T. Miwa, and M. Jimbo, Publ. RIMS, Kyoto Univ. **14**, 223 (1978), **15**, 201 (1979); **15**, 577 (1979); **15**, 871 (1979); **16**, 531 (1980).

23. F. D. M. Haldane, Institut Lane-Langevin Report No. S. P. - 80/126, 1980.

24. V. N. Popov, Freie Universitat, Berlin, Report No. 79-0396, 1979.

25. P. Painlevé, Acta Math. Acad. Sci. Hung. **25**, 1 (1902).

26. L. Schlesinger, J. Reine u. Angew. Math. **141**, 96 (1912).

27. R. Garnier, Ann. Ecol. Norm. Sup. **29**, 1 (1912).

28. T.T.Wu, B.M.McCoy, C.A.Tracy, and E.Barouch, Phys.Rev. B 13, 316 (1976).

29. G. 't Hooft, Nucl.Phys. B 138, 1 (1978).

QUANTUM SPECTRAL TRANSFORM METHOD.
RECENT DEVELOPMENTS.

P.P.Kulish and E.K.Sklyanin

Leningrad Branch of Steklov Mathematical Institute,

Leningrad, USSR

Contents

1. INTRODUCTION

Since quantum electrodynamics has been elaborated the modern quantum field theory developes under the influence of the growing need to go outside the framework of perturbation theory. A striking manifestation of the need is the interest in the exactly soluble models of quantum field theory which increased drastically during the last decade. Though considerable progress has been made only in studying the (1+1)-dimensional field theoretic models [1-4] and the corresponding two-dimensional models of lattice statistics [4-5] , the interest in the subject does not decrease. The reason seems to be, firstly, that even these at the first sight scarcely realistic models can, nevertheless, have a number of common features with the realistic (3+1)-dimensional models. For example, the (1+1)-dimensional nonlinear σ -model is frequently mentioned in the literature to have much in common with the (3+1)-dimensional Yang-Mills theory. On the other hand, solid state physics knows a lot of substances of quasi-one-dimensional structure whose properties are well described by exactly soluble one-dimensional models (Hubbard, XYZ, sine-Gordon models). Finally, quite recently some first promising results have appeared for (2+1)-dimensional case [6, 7] .

The present lectures are devoted to the most promising, from our viewpoint, direction in studying the quantum completely integrable models, the so called Quantum Spectral Transform Method (QSTM, other names are the Quantum Inverse Scattering Method, the Quantum Inverse Problem Method). The QSTM has been developed 3 years ago [8-14] as the result of a synthesis of two major directions in the modern theory of exactly soluble systems. The first of them is based on the tradition of studying the exactly soluble models of solid state and statistical physics which originates from the works of H.Bethe [21] and L.Onsager [22] and culminates in works of R.I.Baxter [23, 24] . The second direction mentioned is the Classical Spectral Transform Method (CSTM), having yielded a lot of important results in the theory of the classical completely integrable systems (for a review see [1-3]). During the last three years a lot of papers on QSTM has been published [8-20, 25-55] and many results have been obtained.

The most impressing achievements of QSTM are:

1. The solution of the sine-Gordon model [13, 20, 26] , i.e. finding its mass spectrum and S-matrix.

2. The solution of the quantum inverse problem for the nonlinear

Schrödinger equation [48, 52] , which enables one to reproduce the known results for the Green's functions of the one-dimensional impenetrable Bose gas. The next problem which is likely to be solved in the nearest future is the calculation of Green's functions for the one-dimensional Bose gas with δ -function interaction.

3. In the purely methodical respect, QSTM has given completeness and transparency to the conventional Bethe ansatz technique. In contrast to the old (coordinate) Bethe ansatz [21, 25, 47] , the new (algebraic) Bethe ansatz [13-15] allows one to obtain the eigenvalues of the Hamiltonian and other integrals of motion immediately omitting the stage which is necessary in the conventional approach, of tediously investigating the eigenfunctions in their manifest coordinate representation.

Intending the present lectures for an audience which is not or is poorly acquainted with QSTM, we did not aim at exhausting description of all results and technicalities (frequently very elaborated ones) of QSTM. We intend merely with use of simple examples to introduce the reader into QSTM and after presenting a general scope of the method and its possibilities, to concentrate on a couple of subjects which are closest to us. The reader interested in more technical details can draw a lot of useful information from the reviews [14-20] and the original papers referred to hereinafter.

We shall now outline the plan of the paper in more detail. Section 2 presents a general survey of QSTM and of the models solved via QSTM. The first two Subsections 2.1. and 2.2 introduce the reader into QSTM using the nonlinear Schrödinger equation as the example. In Subsection 2.3 the QSTM is applied to the massive Thirring model both in Bose and Fermi cases. Subsection 2.4 generalizes QSTM to the lattice models. In Subsection 2.5 a general scheme of applying QSTM is described. Subsection 2.6 contains the list of models which are solved and are to be solved via QSTM.

Sections 3 and 4 deal with the so called Yang-Baxter equation (YBE) which is shown to have many applications in the theory of exactly soluble classical and quantum systems. In Section 3 the origin (Subsection 3.1) and the applications of the Yang-Baxter equation are described. In Section 4 a method of constructing new solutions to the YBE is discussed which is based on the algebraic approach proposed by the authors. The method itself is described in Subsection 4.1, and Subsection 4 2 contains discussion of a new class of exactly soluble models which are generated by the solutions to YBE.

In Section 5 the problem of generalizing the notion of the transition matrix determinant to the quantum case is discussed which arises in applying QSTM to the models including several fields.

Acknowledgements.

The authors are grateful to L.D.Faddeev who has initiated studies on QSTM and to the participants of the Leningrad seminar on the mathematical problems of quantum field theory for many helpful discussions.

2. A GENERAL SURVEY OF QSTM

We begin our review of the basic features of QSTM by considering the nonlinear Schrödinger equation (NS):

$$i\psi_t = -\psi_{xx} + 2\varkappa|\psi|^2\psi\,, \quad \psi(x,t) \underset{|x|\to\infty}{\longrightarrow} 0\,, \tag{2.1}$$

NS is the most appropriate model to begin the acquaintance with QSTM due to the following reasons:

1) The quantum version of NS describes the one-dimensional system of bosons interacting via the δ-function potential. The problem thus reduces to the quantum mechanics of a finite number of particles and does not contain the difficulties of relativistic quantum field theory (non-Fock representations of CCR, divergences etc.).

2) NS has been studied in detail both in classical and quantum cases. The classical NS admits application of CSTM [1] . In the quantum case there exists a complete description of the spectrum and the eigenfunctions of the Hamiltonian [56-58] . The latter circumstance allows one to compare the new results with the already known ones.

Before we shall deal in detail with NS let us compare the statement of the problem in classical and quantum cases. In classical mechanics, one is interested usually in the evolution of the initial data and CSTM is used to find the time-evolution of the scattering data for an auxiliary linear problem. For quantum mechanics, on the contrary, the "stationary" approach is more characteristic which deals with the spectrum of the Hamiltonian and S-matrix. However, in spite of the apparent unlikeness of the two statements of the problem there exists an approach to CSTM which is quite similar to the "stationary" quantum mechanical one. The approach referred to is the Hamiltonian interpretation of CSTM developed first in [59] . The final aim of investigation of a nonlinear evolution equation within this approach is to construct the action-angle variables from the scattering data of the auxiliary linear problem for the nonlinear equation and to prove thus its complete integrability. As a by-product one obtains the spectrum of elementary excitations. It is the Hamiltonian approach to CSTM which we shall take as the starting point for the quantum generalization.

2.1. CSTM for the Nonlinear Schrödinger equation

We recall some basic stages of CSTM as applied to NS. The way of presentation we choose here may seem a little bit unusual but it is the most appropriate one for subsequent quantum generalizations.

The basic object of CSTM is the so-calledtransition (or monodromy) matrix $T(x_1, x_2, \lambda)$ which is defined as the solution to the initial value problem:

$$\frac{\partial}{\partial x_2} T(x_1, x_2; \lambda) = L(x_2, \lambda) T(x_1, x_2; \lambda); \qquad T(x, x; \lambda) = I_2 \tag{2.2}$$

where

$$L(x, \lambda) = i \begin{pmatrix} -\frac{\lambda}{2} & æ\bar{\psi}(x) \\ -\psi(x) & \frac{\lambda}{2} \end{pmatrix}; \quad I_2 = \begin{pmatrix} 1 & 0 \\ 0 & 1 \end{pmatrix}. \tag{2.3}$$

The transition matrix $T(\lambda)$ for the infinite interval $(-\infty, \infty)$ is obtained from $T(x_1, x_2, \lambda)$ in the limit $x_1 \to -\infty$, $x_2 \to +\infty$ after the proper renormalization:

$$T(\lambda) = \lim_{\substack{x_1 \to -\infty \\ x_2 \to +\infty}} e^{i\frac{\lambda}{2}\sigma_3 x_2} T(x_1, x_2; \lambda) e^{-i\frac{\lambda}{2}\sigma_3 x_1}; \quad \sigma_3 = \begin{pmatrix} 1 & 0 \\ 0 & -1 \end{pmatrix} \tag{2.4}$$

Due to the symmetries of the L-operator (2.3) the matrix T looks like

$$T(\lambda) = \begin{pmatrix} a(\lambda) & æ\bar{b}(\lambda) \\ b(\lambda) & \overline{a(\lambda)} \end{pmatrix}, \quad \lambda = \bar{\lambda}. \tag{2.5}$$

The matrix element $a(\lambda)$ ($\bar{a}(\lambda)$) admits analytical continuation into upper (lower) halfplane of the spectral parameter λ.

An important point is that NS (2.1) can be considered as a Hamiltonian system with the Hamiltonian

$$H = \int dx \left(|\psi_x|^2 + æ|\psi|^4 \right) \tag{2.6}$$

and canonical Poisson brackets

$$\{\psi(x), \psi(y)\} = \{\bar{\psi}(x), \bar{\psi}(y)\} = 0, \quad \{\psi(x), \bar{\psi}(y)\} = i\delta(x-y) \tag{2.7}$$

so that

$$\psi_t = \{H, \psi\} \tag{2.8}$$

The next step of CSTM consists in calculating the Poisson brackets between the quantities $a(\lambda)$, $b(\lambda)$, $\bar{a}(\lambda)$, $\bar{b}(\lambda)$ (2.5) considered as functionals of the dynamical variables $\psi(x)$, $\bar{\psi}(x)$. They turn out to be [8, 9, 11, 15, 18]

$$\begin{aligned}
&\{a(\lambda), a(\mu)\} = \{a(\lambda), \bar{a}(\mu)\} = \{b(\lambda), b(\mu)\} = 0, \\
&\{a(\lambda), b(\mu)\} = \frac{\varkappa}{\lambda-\mu+i0}\, a(\lambda) b(\mu), \\
&\{a(\lambda), \bar{b}(\mu)\} = \frac{-\varkappa}{\lambda-\mu+i0}\, a(\lambda)\bar{b}(\mu), \\
&\{b(\lambda), \bar{b}(\mu)\} = 2\pi i\, |a(\lambda)|^2 \delta(\lambda-\mu).
\end{aligned} \tag{2.9}$$

To derive (2.9) following an idea of [9] we introduce first some notation which frequently will be used hereafter. For arbitrary 2×2 matrix $T = \begin{pmatrix} t_{11} & t_{12} \\ t_{21} & t_{22} \end{pmatrix}$ we define two 4×4 matrices T' and T'' by

$$T' = T\otimes I_2 = \begin{pmatrix} t_{11} & 0 & t_{12} & 0 \\ 0 & t_{11} & 0 & t_{12} \\ t_{21} & 0 & t_{22} & 0 \\ 0 & t_{21} & 0 & t_{22} \end{pmatrix}, \quad T'' = I_2 \otimes T = \begin{pmatrix} t_{11} & t_{12} & 0 & 0 \\ t_{21} & t_{22} & 0 & 0 \\ 0 & 0 & t_{11} & t_{12} \\ 0 & 0 & t_{21} & t_{22} \end{pmatrix} \tag{2.10}$$

Using the above notation we can write down all the 16 Poisson brackets between the matrix elements of $T(\lambda)$ and $T(\mu)$ in compact form $\{T'(\lambda), T''(\mu)\}$ as a 4×4 matrix.

The following relation which is verified directly is crucial for the whole method:

$$\{L'(x,\lambda), L''(y,\mu)\} = [r(\lambda-\mu), L'(x,\lambda) + L''(y,\mu)]\,\delta(x-y) \tag{2.11}$$

where

$$r(\lambda) = -\frac{\varkappa}{\lambda}\begin{pmatrix} 1 & & & \\ & 0 & 1 & \\ & 1 & 0 & \\ & & & 1 \end{pmatrix} \tag{2.12}$$

From the local equality (2.11) the global relation for $T(x_1, x_2; \lambda)$:

$$\{T'(x_1,x_2;\lambda), T''(x_1,x_2;\lambda)\} = [r(\lambda-\mu), T'(x_1,x_2;\lambda)T''(x_1,x_2;\mu)] \qquad (2.13)$$

can be easily derived.

Passing carefully to the limit $x_1 \to -\infty$, $x_2 \to +\infty$ in (2.13) we obtain [18]

$$\{T'(\lambda), T''(\mu)\} = r_+(\lambda-\mu)T'(\lambda)T''(\mu) - T'(\lambda)T''(\mu)r_-(\lambda-\mu) \qquad (2.14)$$

where

$$r_\pm(\lambda-\mu) = -\varkappa \begin{pmatrix} \mathrm{v.p.}\frac{1}{\lambda-\mu} & 0 & 0 & 0 \\ 0 & 0 & \pm i\pi\delta(\lambda-\mu) & 0 \\ 0 & \mp i\pi\delta(\lambda-\mu) & 0 & 0 \\ 0 & 0 & 0 & \mathrm{v.p.}\frac{1}{\lambda-\mu} \end{pmatrix} \qquad (2.15)$$

The Poisson brackets (2.9) are contained in (2.14). The method presented is not probably the shortest possible one, however, as we shall see further, it admits a direct quantum generalization.

Let us now proceed to the interpretation of the scattering data (2.5) and the Poisson brackets (2.9). It is well known [1] that $\ln a(\lambda)$ can be expanded in powers of λ^{-1}

$$\ln a(\lambda) = \sum_{m=1}^{\infty} J_m \lambda^{-m} \qquad (2.16)$$

The coefficients J_m are integrals of local densities in $\psi(x)$, $\bar\psi(x)$. The first three coefficients J_1, J_2, J_3 turn out to be the number of particles N, momentum P and the Hamiltonian H (2.6) resp.

$$N = \int |\psi|^2 dx \qquad (2.17)$$

$$P = \frac{i}{2}\int (\bar\psi_x \psi - \bar\psi \psi_x)\, dx \qquad (2.18)$$

It follows from $\{a(\lambda), a(\mu)\} = 0$ (2.9) that $\{J_m, J_n\} = 0$ so that J_m form an infinite set of integrals of motion for (2.1).

Due to (2.9) the quantities $\varphi(\lambda)$, $\bar\varphi(\mu)$ given by

$$\varphi(\lambda) = \frac{b(\lambda)}{|b(\lambda)|}\sqrt{\frac{\ln|a(\lambda)|}{\pi \varkappa}} \tag{2.19}$$

have the canonical Poisson brackets

$$\begin{aligned} &\{\varphi(\lambda), \varphi(\mu)\} = \{\bar{\varphi}(\lambda), \bar{\varphi}(\mu)\} = 0, \\ &\{\varphi(\lambda), \bar{\varphi}(\mu)\} = i\delta(\lambda-\mu) \end{aligned} \tag{2.20}$$

The generating function $\ln a(\lambda)$ of the integrals of motion written down in terms of $\varphi(\lambda)$, $\bar{\varphi}(\lambda)$ is $(\varkappa > 0)$

$$\ln a(\lambda) = \frac{i}{\pi}\int_{-\infty}^{\infty} d\mu \frac{\ln|a(\mu)|}{\lambda-\mu} = i\varkappa \int_{-\infty}^{\infty} d\mu \frac{|\varphi(\mu)|^2}{\lambda-\mu} \tag{2.21}$$

which implies

$$J_m = \int_{-\infty}^{\infty} \mu^{m-1} |\varphi(\mu)|^2 d\mu \tag{2.22}$$

Thus, all J_m are quadratic in φ , $\bar{\varphi}$ and the corresponding equations of motion

$$\varphi_t = \{J_m, \varphi\} = -i\mu^{m-1}\varphi \tag{2.23}$$

are linear, the fact which allows us to consider the variables φ , $\bar{\varphi}$ as the action-angle variables for NS (2.1).

To conclude our discussion of CSTM we shall point out the basic distinction between our variant of CSTM and the conventional one (see for example [1-3]).

The conventional CSTM is based on representing a given nonlinear evolution equation as the compatibility condition

$$L_t = M_x + [M, L] \tag{2.24}$$

of the set of two equations

$$\begin{cases} T_x = LT \\ T_t = MT \end{cases} \tag{2.25}$$

In our "stationary" approach we use only the first of two eqns.(2.25) and the Hamiltonian structure. The role of the criterion of complete integrability is played now by the eqn.(2.11). Thus the arsenal of CSTM is enriched with a new powerful tool, the r-matrix. Though, at present, there is no deep theoretical reason for r-matrix to exist and the validity of the equality (2.11) for any new model is rather a matter of fortune, the intriguing fact is that the r-matrix approach can be extended to a broad variety of completely integrable models. Among them one should mention sine-Gordon equation, Shabat--Mikhailov model [27] , Heisenberg ferromagnets [60] . The r -matrices for these models are of course different from (2.12). All the models listed above have one common feature: they are ultralocal (in the sense of [15]), i.e. the Poisson brackets of the matrix elements of the L-operator do not contain any derivatives of δ -function, which is manifestly assumed in (2.11). It was shown in [45] that for nonultralocal models such as Korteweg-de-Vries eqn., sine-Gordon in the light cone variables and someothers the r -matrix method is also applicable. In the nonultralocal case the eqn. (2.11) certainly must be somehow generalized but the eqn. (2.13) remains valid. Throughout the present paper we consider only the ultralocal case.

2.2. QSTM for the quantum NS

In the quantum case the dynamical variables $\psi(x)$, $\bar{\psi}(x)$ are replaced by the annihilation and creation operators $\Psi(x)$, $\Psi^+(x)$

$$[\Psi(x),\Psi(y)]=[\Psi^+(x),\Psi^+(y)]=0,\quad [\Psi(x),\Psi^+(y)]=\delta(x-y) \tag{2.26}$$

The essence of QSTM consists in formulating the problem of quantum generalization of CSTM as the problem of constructing the quantum operators $A(\lambda)$, $A^+(\lambda)$, $B(\lambda)$, $B^+(\lambda)$ possessing, by analogy with CSTM, the following properties [8] .

1. As in CSTM, $\ln A(\lambda)$ is a generating function of mutually commuting conserved quantities.

2. The operators $B(\lambda)$, $B^+(\lambda)$ are the quantum analogues of the action-angle variables, i.e. annihilate and create the excitations of the system.

The solution of the problem stated is quite simple (for NS only !). The operators $A(\lambda)$, $A^+(\lambda)$, $B(\lambda)$, $B^+(\lambda)$ are obtained from a(λ), $\bar{a}$(λ), b(λ), $\bar{b}(\lambda)$ after replacing ψ , $\bar{\psi}$ by Ψ , Ψ^+ and subsequent normal ordering (in other words, the functionals a(λ) etc. are Wick symbols of the operators A(λ) etc.). This connection can be expressed by

$$A(\lambda)=:a(\lambda):,\ A^+(\lambda)=:\bar{a}(\lambda):,\ B(\lambda)=:b(\lambda):,\ B^+(\lambda)=:\bar{b}(\lambda):. \tag{2.27}$$

The commutation relations between the operators A, A^+, B, B^+ turn out to be [8-10, 15, 18] :

$$\begin{gathered}
[A(\lambda),A(\mu)]=[A(\lambda),A^+(\mu)]=[B(\lambda),B(\mu)]=0,\\
B(\mu)A(\lambda)=\left(1+\frac{i\varkappa}{\lambda-\mu+i0}\right)A(\lambda)B(\mu),\\
A(\lambda)B^+(\mu)=\left(1+\frac{i\varkappa}{\lambda-\mu+i0}\right)B^+(\mu)A(\lambda),\\
B(\lambda)B^+(\mu)=\left(1+\frac{i\varkappa}{\lambda-\mu+i0}\right)\left(1-\frac{i\varkappa}{\lambda-\mu-i0}\right)B^+(\mu)B(\lambda)+2\pi A^+(\lambda)A(\lambda)\delta(\lambda-\mu)
\end{gathered} \tag{2.28}$$

It follows from (2.28) that $\ln A(\lambda)$ is a commutative family of operators. Moreover, $\ln A(\lambda)$ can be shown [18] to be the generating function for mutually commuting operators including the number of particles N, momentum P and Hamiltonian H. The operator $\ln A(\lambda)$ commutes with $B^+(\mu)$ as follows

$$[\ln A(\lambda),B^+(\mu)]=B^+(\mu)\ln\frac{\lambda-\mu+i\varkappa}{\lambda-\mu+i0} \tag{2.29}$$

The general N-particle state is obtained by applying B^+ N-times to the ground state $|o\rangle$:

$$|k_1,\dots,k_N\rangle=B^+(k_1)\dots B^+(k_N)|0\rangle. \tag{2.30}$$

The wave function of the state $|k_1,\dots,k_N\rangle$ turns out to coincide with the wave function obtained by Bethe ansatz [56-58] .

The eigenvalues of $\ln A(\lambda)$ are obtained from (2.29):

$$\begin{gathered}
\ln A(\lambda)|k_1,\dots,k_N\rangle=\sum_{j=1}^{H}\ln\frac{\lambda-k_j+i\varkappa}{\lambda-k_j}|k_1,\dots,k_N\rangle,\\
\hat{N}|k_1,\dots,k_N\rangle=N|k_1,\dots,k_N\rangle,\\
\hat{P}|k_1,\dots,k_N\rangle=\sum_{j=1}^{N}k_j|k_1,\dots,k_N\rangle,
\end{gathered} \tag{2.31}$$

$$\hat{H}\,|k_1,\dots,k_N\rangle = \sum_{j=1}^{N} k_j^2\,|k_1,\dots,k_N\rangle$$

The equalities (2.30-2.31) allow one to interprete $B^+(\lambda)$ $(B(\lambda))$ as a creation (annihilation) operator for one particle carrying the momentum λ and energy λ^2 . Strictly speaking, B and B^+ are not canonically commuting operators (cf. (2.28)). This disadvantage, however, can be easily removed by renormalizing B's. In fact, the operators Φ and Φ^+

$$\Phi(\lambda) = \left(2\pi A^+(\lambda)A(\lambda)\right)^{-1/2} B(\lambda)$$

$$\Phi^+(\lambda) = B^+(\lambda)\left(2\pi A^+(\lambda)A(\lambda)\right)^{-1/2} \qquad (2.32)$$

as a consequence of (2.28) satisfy the canonical commutation relations:

$$[\Phi(\lambda),\Phi(\mu)] = [\Phi^+(\lambda),\Phi^+(\mu)] = 0, \quad [\Phi(\lambda),\Phi^+(\mu)] = \delta(\lambda-\mu) \qquad (2.33)$$

Let us describe now the method of deriving the commutation relations (2.28). As in the classical case, it is useful to start from the finite interval and to deal with the quantum transition matrix $\hat{T}(x_1, x_2; \lambda)$ which is defined as the normally ordered classical matrix $T(x_1, x_2; \lambda)$:

$$\hat{T}(x_1,x_2;\lambda) = :T(x_1,x_2;\lambda):\,, \quad \hat{T} = \begin{pmatrix} A & \text{æ} B^+ \\ B & A^+ \end{pmatrix} \qquad (2.34)$$

The differential equation for $\hat{T}(x_1, x_2; \lambda)$ is obtained from (2.2) by normal ordering:

$$\frac{\partial}{\partial x_2}\hat{T}(x_1,x_2;\lambda) = :\hat{L}(x_2,\lambda)\hat{T}(x_1,x_2;\lambda): \qquad (2.35)$$

The following fundamental relation

$$R(\lambda-\mu)\,\hat{T}'(x_1,x_2;\lambda)\,\hat{T}''(x_1,x_2;\mu) = \hat{T}''(x_1,x_2,\mu)\,\hat{T}'(x_1,x_2;\lambda)\,R(\lambda-\mu) \qquad (2.36)$$

where

$$R(\lambda) = 1 - i\frac{\text{æ}}{\lambda}\begin{pmatrix} 1 & & & \\ & 0 & 1 & \\ & 1 & 0 & \\ & & & 1 \end{pmatrix}$$

plays in QSTM the same role as (2.13) does in CSTM. Thus a new essential for QSTM object arises, the R-matrix.

It is worth noticing that (2.13) is obtained from (2.36) in the classical limit $\hbar \to 0$, assuming the coupling constant $\varkappa$ to be proportional to the Planck constant $\hbar$, inserting into (2.36) the expansion

$$R(\lambda) = 1 + i\hbar r(\lambda) + O(\hbar^2) \tag{2.37}$$

using the correspondence principle $[\ ,\ ] \sim -i\hbar\{\ ,\ \}$ and taking from (2.36) the terms of order $\hbar$.

As in the classical case, to prove (2.36) it is enough to verify the local analog of (2.36) which is obtained by differentiating (2.36) with respect to x_2 at $x_1 = x_2$ [18] :

$$\begin{aligned} &R(\lambda-\mu)\left(\hat{L}'(x_2,\lambda) + \hat{L}''(x_2,\mu) + \varkappa\sigma'_-\sigma''_+\right) = \\ &= \left(\hat{L}'(x_2,\lambda) + \hat{L}''(x_2,\mu) + \varkappa\sigma'_+\sigma''_-\right) R(\lambda-\mu) \end{aligned} \tag{2.38}$$

where $\sigma_- = \begin{pmatrix} 0 & 0 \\ 1 & 0 \end{pmatrix}$, $\sigma_+ = \begin{pmatrix} 0 & 1 \\ 0 & 0 \end{pmatrix}$.

The terms $\varkappa\sigma'_-\sigma''_+$ and $\varkappa\sigma'_+\sigma''_-$ in (2.38) are the quantum corrections which arise due to the noncommutativity of Ψ and Ψ^+.

The commutation relations (2.38) for the infinite interval can be obtained from (2.36) in the limit $x_1 \to -\infty$, $x_2 \to +\infty$. The limiting process must be performed with care and involves a renormalization procedure which we do not describe here (see [15, 18]).

It is to be noted that all the above reasoning is based on the crucial fact already mentioned in the very beginning of the Section 2. Namely, the representation of CCR (2.26) for the interacting field $\Psi(x)$, $\Psi^+(x)$ coincides with that for the free field. Firstly, we used it in defining A, A^+, B, B^+ as normally ordered operators. Secondly the relations $B(\lambda)|0\rangle = 0$ and $-[B^+(\lambda), \hat{N}] = B^+(\lambda)$ which allow to interprete B and B^+ as the annihilation and creation operators resp. are inherited from the corresponding properties of Ψ and Ψ^+ which are characteristic of the Fock representation of CCR.

Of great physical interest, however, are the theories in which non-Fock representations arise. The simplest example is the Bose-gas of finite density. In classical case it is described by NS equation with the boundary condition $|\psi| \to \text{const}$, $|x| \to \infty$. In the quantum case one needs to modify the Hamiltonian $\hat{H}_\mu = \hat{H} - \mu\hat{N}$, $\hat{H}$ being old NS Hamiltonian and μ being the chemical potential. Apparently the CCR representation for Ψ, Ψ^+ is still not a priori known and must have very complicated structure. An attempt to consider $\hat{H}_\mu$ in the Fock representation space $\mathcal{H}_F$ leads to the fact that $\hat{H}_\mu$ is unbounded from below.

A method of solving the problem has been proposed by Lieb and Liniger [58] who used the Bethe ansatz technique. They put the system in the box of length $2l$ with the periodic boundary conditions. After the space cutoff is introduced the Fock CCR representation can be used, since the space $\mathcal{H}_F$ contains now both the Fock vacuum $|0\rangle$ and the physical vacuum Ω which minimizes $\hat{H}_\mu$. The latter statement follows from the fact that energy difference between $|0\rangle$ and Ω is now finite.

Since Ω is contained in the Fock space $\mathcal{H}_F$ built over $|0\rangle$ one can look for the state Ω in the form

$$\Omega = B^+(k_1) \dots B^+(k_N)|0\rangle, \tag{2.39}$$

k_j being unknown. To proceed, it is more convenient to minimize H at fixed value of N rather than to minimize $\hat{H} - \mu\hat{N}$. It is well known that in the thermodynamical limit both procedures are equivalent.

The set of transcendental equations for momenta $\{k_j\}_1^N$ has been found in [58] from the periodicity conditions for the wave-function constructed by means of the Bethe ansatz. Here, we obtain the same equations in frame of the QSTM, following [15] .

Note, first of all, that due to invertibility of the R-matrix (2.37) in (2.39) the 4×4 -matrices $\hat{T}'(\lambda)\,\hat{T}''(\mu)$ and $\hat{T}''(\mu)\,\hat{T}'(\lambda)$ are similar. On multiplying (2.36) by $R^{-1}(\lambda-\mu)$ and taking the matrix trace we arrive at the important corollary. The matrix traces of the transition matrices commute (being considered as quantum-mechanical operators in $\mathcal{H}_F$).

$$[t(\lambda), t(\mu)] = 0, \quad t(\lambda) = \mathrm{tr}\,\hat{T}(\lambda) = A(\lambda) + A^+(\lambda), \tag{2.40}$$
$$\mathrm{tr}\,\hat{T}'(\lambda)\,\hat{T}''(\mu) = \mathrm{tr}\,\hat{T}(\lambda)\,\mathrm{tr}\,\hat{T}(\mu)$$

The operator family $t(\lambda)$ plays for the finite interval $[x_1, x_2]$ the same role of generating function of integrals of motion which is played for the infinite interval $(-\infty,\infty)$ by $A(\lambda)$. So, we can get the periodicity equations for k_j from the condition that $|k_1 \dots k_N\rangle$ is an eigenvector of $t(\lambda)$.

To this end, let us write down explicitly some of the relations (2.36)

$$A(\lambda)B^+(\mu)=B^+(\mu)A(\lambda)\frac{\lambda-\mu+i\varkappa}{\lambda-\mu}-\frac{i\varkappa}{\lambda-\mu}B^+(\lambda)A(\mu), \tag{2.41}$$

$$A^+(\lambda)B^+(\mu)=B^+(\mu)A^+(\lambda)\frac{\lambda-\mu-i\varkappa}{\lambda-\mu}+\frac{i\varkappa}{\lambda-\mu}B^+(\lambda)A^+(\mu), \tag{2.42}$$

$$B^+(\lambda)B^+(\mu)=B^+(\mu)B^+(\lambda) \tag{2.43}$$

Note that the Fock vacuum is an eigenvector of both operators $A(\lambda)$ and $A^+(\lambda)$

$$A(\lambda)|0> = e^{-i\lambda\ell}|0>, \quad A^+(\lambda)|0> = e^{i\lambda\ell}|0>, \quad 2\ell = x_2 - x_1 \tag{2.44}$$

Consider now the expression

$$t(\lambda)|k_1,\dots,k_N> = \left(A(\lambda)+A^+(\lambda)\right)B^+(k_1)\dots B^+(k_N)|0> \tag{2.45}$$

To obtain the equations needed let us carry $A(\lambda)$ and $A^+(\lambda)$ in (2.45) through all the B^+'s using (2.41, 42, 44). The result is

$$t(\lambda)|k_1,\dots,k_N> = \nu(\lambda;k_1,\dots,k_N)|k_1,\dots,k_N> + \tag{2.46}$$
$$+\sum_{j=1}^{N}\Lambda_j(\lambda;k_1,\dots,k_N)B^+(\lambda)\prod_{\substack{i\neq j\\ i=1}}^{N}B^+(k_i)|0>$$

where

$$\nu(\lambda;k_1,\dots,k_N) = e^{-i\lambda\ell}\prod_{j=1}^{N}\frac{\lambda-k_j+i\varkappa}{\lambda-k_j} + e^{i\lambda\ell}\prod_{j=1}^{N}\frac{\lambda-k_j-i\varkappa}{\lambda-k_j} \tag{2.47}$$

The first summand in the r.h.s. of (2.46) is obtained when we carry $A(\lambda)+A^+(\lambda)$ through B^+'s using the first terms of the r.h.s. of (2.41-42) only. The second summand contains all the remaining "unwanted" terms which must vanish after imposing the condition that $|k_1,\dots,k_N>$ be an eigenvector of $t(\lambda)$. We do not need to write the explicit expression for $\Lambda_j(\lambda;k_1\dots k_N)$ because, as S.V.Manakov has noticed [61] the condition $\Lambda_j=0$ is equivalent to the requirement that the eigenvalue $\nu(\lambda;k_1,\dots$ $\dots,k_N)$ has no singularities in λ. Thus, from $\mathrm{Res}\,\nu(\lambda;k_1,\dots,k_N)=0$ for $\lambda=k_j$ it follows

$$e^{-2ik_j l} = \prod_{\substack{i=1 \\ i\neq j}}^{N} \frac{k_j - k_i - i\varkappa}{k_j - k_i + i\varkappa}, \quad j = 1,2,\dots,N \tag{2.48}$$

The periodicity equation (2.48) lays the basis of the further investigation of the ground state and elementary excitations of the Bose-gas with finite density in the thermodynamical limit ($l \to \infty$, N/l being constant) [58]. The reader can find details in the original literature cited.

2.3. QSTM for the massive Thirring model

We proceed now to describe some modifications of QSTM for models more complicated than NS. It is instructive to observe how the simple scheme described above gradually becomes more and more complicated.

The simplest model after NS is the massive Thirring (MT) model which describes the relativistic two-dimensional massive spinor field with current-current interaction. The Lagrangian $\mathcal{L}$ and the Hamiltonian H of the model are

$$\mathcal{L} = \bar{\psi}(i\gamma^\mu \partial_\mu - m)\psi - \tfrac{1}{2} g j^\mu j_\mu, \quad j^\mu = \bar{\psi}\gamma^\mu\psi, \tag{2.49}$$

$$\gamma^0 = \begin{pmatrix} 0 & 1 \\ 1 & 0 \end{pmatrix}, \quad \gamma^1 = \begin{pmatrix} 0 & -1 \\ 1 & 0 \end{pmatrix}, \quad \gamma^5 = \gamma^0\gamma^1, \quad \psi = \begin{pmatrix} \psi_1 \\ \psi_2 \end{pmatrix},$$

$$H = \int dx \left(\psi^+(-i\gamma^5\partial_1 + \gamma^0 m)\psi + 2g\,\psi_1^+\psi_2^+\psi_2\psi_1 \right). \tag{2.50}$$

The classical variant of MT is soluble via CSTM both in cases of commuting and anticommuting (Grassmann) variables ψ , $\bar{\psi}$. In the commuting case the L-operator is [62] 2×2 matrix ($\rho_a = \psi_a^+\psi_a$)

$$L(x,\lambda) = i \begin{pmatrix} \dfrac{-\lambda^2+\lambda^{-2}}{4} + \rho_1 - \rho_2 & \lambda\psi_2^+ - \dfrac{1}{\lambda}\psi_1^+ \\ \lambda\psi_2 - \dfrac{1}{\lambda}\psi_1 & \dfrac{\lambda^2-\lambda^{-2}}{4} - \rho_1 + \rho_2 \end{pmatrix}, \quad m = g = 1 \tag{2.51}$$

whereas in the anticommuting case it is 3×3 matrix [63]

$$L(x,\lambda)=-i\begin{pmatrix} \rho_1-\rho_2 & 0 & \lambda\psi_1+\frac{1}{\lambda}\psi_2 \\ 0 & -\rho_1+\rho_2 & -\lambda\psi_1^+ + \frac{1}{\lambda}\psi_2^+ \\ \lambda\psi_1^+ + \frac{1}{\lambda}\psi_2^+ & -\lambda\psi_1+\frac{1}{\lambda}\psi_2 & \frac{-\lambda^2+\lambda^{-2}}{2} \end{pmatrix}. \quad (2.52)$$

The main difficulty of dealing with the quantum fermionic MT is the same as in the case of NS with finite density. The ACR representation for the physical vacuum Ω is a priori unknown whereas over the nonphysical vacuum $|0\rangle$ (pseudovacuum) the spectrum of H (2.50) is unbounded from below. The excitations over $|0\rangle$ (pseudoparticles) can be described in terms of Bethe ansatz technique [64] and the physical vacuum Ω can be constructed in the manner very much as in the case of NS introducing spatial and ultraviolet cutoffs and filling the Dirac sea of pseudoparticles with negative energies. The integral equations for pseudoparticle densities, the spectrum of physical excitations and their S-matrices were calculated within the Bethe ansatz approach in [25, 26, 47].

In the Bose case only nonphysical Fock representation of CCR exists for MT but, nevertheless, the Bose-MT is a useful toy for mastering QSTM.

The attempt to algebraize the Bethe ansatz for MT along the same lines as in the case of NS leads immediately to a failure. If the quantum transition matrix $T(x_1, x_2; \lambda)$ is defined by the formulas (2.34, 35) using the classical L operators (2.51) or (2.52) there is no such R-matrix for (2.36) to be valid. The important lesson one must learn from this failure is that in general case the quantum L-operator must differ from the classical one coinciding with it in the classical limit only. A naive treatment of the normally ordered reflection coefficient $\bar{b}(\lambda)$ as a creation operator [12] does not lead to the right answer for the eigenfunctions.

The quantum L-operators for the Bose- and Fermi-cases are, respectively,

$$L(x,\lambda)=i\begin{pmatrix} \frac{-\lambda^2+\lambda^{-2}}{4}-\xi\rho_1+\eta\rho_2 & \lambda\psi_2^+ - \frac{1}{\lambda}\psi_1^+ \\ \lambda\psi_2-\frac{1}{\lambda}\psi_1 & \frac{\lambda^2-\lambda^{-2}}{4}+\eta\rho_1-\xi\rho_2 \end{pmatrix}, \quad (2.53)$$

$$\eta=e^{2\gamma}\xi=e^{\gamma}/\cosh\gamma,\quad \gamma=\frac{i}{2}\arcsin\hbar,\quad [\psi_m(x),\psi_n^+(y)]=\hbar\,\delta_{mn}\,\delta(x-y)$$

and

$$L(x,\lambda)=-i\begin{pmatrix} \xi\rho_1-\eta\rho_2 & 0 & \lambda\psi_1+\frac{1}{\lambda}\psi_2 \\ 0 & \tau\lambda^2+\theta\frac{1}{\lambda^2}-\eta\rho_1+\xi\rho_2 & -\alpha\lambda\psi_1^+ +\frac{1}{\alpha\lambda}\psi_2^+ \\ \lambda\psi_1^+ +\frac{1}{\lambda}\psi_2^+ & -\alpha\lambda\psi_1+\frac{1}{\alpha\lambda}\psi_2 & \frac{-\lambda^2+\lambda^{-2}}{2} \end{pmatrix} \qquad (2.54)$$

$$\alpha=e^{\gamma},\ \eta=\alpha^2\xi=e^{\gamma}/\cosh\gamma,\ \tau=\alpha^2\theta=e^{\gamma}\sinh\gamma,\ \gamma=\frac{i}{2}\operatorname{arcsin}\hbar,\ \{\psi_m(x),\psi_n^+(y)\}=\delta_{nm}\hbar\,\delta(x-y)$$

and the corresponding R -matrices are

$$R(u,\gamma)=\left|\begin{matrix} a & & & \\ & b & c & \\ & c & b & \\ & & & a \end{matrix}\right| \quad \begin{aligned} & a=1, \qquad \exp u=\lambda/\mu, \\ & b=\sinh u/\sinh(u+2\gamma), \\ & c=\sinh 2\gamma/\sinh(u+2\gamma) \end{aligned} \qquad (2.55)$$

and [19]

$$R(u,\gamma)=\left|\begin{array}{ccc|ccc|ccc} a & & & & & & & & \\ & b & & z & & & & & y \\ & & c & & & & x & & \\ \hline & z & & b & & & & & y \\ & & & & a & & & & \\ & & & & & c & & x & \\ \hline & & x & & & & c & & \\ & & & & & x & & c & \\ & y & & y & & & & & d \end{array}\right| \quad \begin{aligned} & a=1, \quad \exp u=\lambda/\mu, \\ & b=J\sinh u\cosh(u-\gamma)/\cosh(u+\gamma), \\ & z=J\sinh 2\gamma\sinh\gamma/\cosh(u+\gamma), \\ & c=J\sinh u, \\ & d=c-z, \\ & x=J\sinh 2\gamma, \\ & y=J\sinh u\sinh 2\gamma/\cosh(u+\gamma), \\ & J=1/\sinh(u+2\gamma). \end{aligned} \qquad (2.56)$$

In the Fermi case the fundamental equation (2.36) must be understood in the graded sense [19].

In the Bose case one can follow the same line as for NS, obtaining the generating function of the local commuting integrals of motion from the diagonal elements of T(λ) and the creation annihilation operators for (nonphysical) excitations over $|0\rangle$ from the off-diagonal elements.

Unfortunately, in the Fermi case which is of direct physical interest the situation is more complicated. There are too many matrix elements in the 3×3 matrix T(λ) and the commutation relations between them are too complicated. The problem of constructing the creation annihilation operators for the Fermi MT remains still unsolved.

2.4. QSTM for sine-Gordon and lattice models

The above examples illustrate well the fundamental significance of the pseudovacuum $|o\rangle$ for QSTM. Unfortunately, in the relativistic quantum field theories containing Bose fields,such as sine-Gordon (SG), supersymmetric sine-Gordon (SSG), Shabat-Mikhailov model (SM), there is no a priori known CCR representation for the interacting fields containing the reference state $|o\rangle$ which could play role of pseudovacuum. In case of SG model the difficulty can be, of course, avoided using the Coleman's equivalence of SG and MT [65]. This way, however, can scarcely be considered as an honest one.

Another problem which arises in dealing with the relativistic fields is the ultraviolet renormalization. In case of MT, the usual way of removing the divergencies [25, 26, 47] consists of introducing a cutoff or, in other words, of filling the Dirac sea up to the bound rapidity Λ. Introducing Λ seems to be somewhat artificial and is not dictated by the periodicity equations of (2.48) type *).

A remedy against the difficulties indicated above is provided by putting the system on the lattice. The lattice spacing Δ provides a natural intrinsic scale for the ultraviolet cutoff and the continuum field theory is obtained by tending $\Delta \to 0$ and the coupling constants of the lattice Hamiltonian to a critical point [66].

A single continuous field model, however, has a lot of lattice versions. The question arises if there are exactly soluble ones among them. Fortunately, for most interesting models (NS, SG) the answer turns out to be positive (and we believe it to be positive in general case) and, what is more, there can be several exactly soluble lattice approximations for a single continuous field model (for NS, there are at least 3).

To proceed, let us describe the main features of QSTM and CSTM in the lattice case. The field operators are defined on the lattice $x=o, \pm\Delta, \pm 2\Delta, \ldots$. The fundamental relations (2.13) and (2.36) for classical and quantum cases respectively do not change their form. The L -operator $L(x, \lambda)$ is defined simply as the transition matrix $T(x, x+\Delta; \lambda)$ for one step, so that

---

*) The difficulties with the renormalization procedure in the continuous approach were underlined also by B.M.McCoy [private communication].

$$T(x_1, x_2; \lambda) = L(x_2-\Delta;\lambda)\ldots L(x_1;\lambda) \tag{2.57}$$

both in classical and quantum cases. It is to be noted that the problem of normal ordering is completely removed being on the lattice.

The local identities, like (2.11), which generate (2.13) and (2.36) read

$$\{L'(x,\lambda), L''(x,\mu)\} = [\, r(\lambda-\mu), L'(x,\lambda)L''(x,\mu)\,] \tag{2.58}$$

in classical case, and

$$R(\lambda-\mu)L'(x,\lambda)L''(x,\mu) = L''(x,\mu)L'(x,\lambda)R(\lambda-\mu) \tag{2.59}$$

in quantum case. We would remind the reader that throughout the present paper we consider the ultralocal case. The condition of ultralocality for the lattice models means that the field operators at different sites of the lattice must commute and $L(x, \lambda)$ contains only the field operators at the site x.

Though the lattice approach is motivated mainly by studying the relativistic field models, we shall describe here, for the sake of simplicity, the completely integrable lattice versions of the NS model.

1. XXZ model is formulated in terms of spin - $\frac{1}{2}$ operators

$$S_m^\alpha S_n^\beta = \begin{cases} \delta_{\alpha\beta} + i\varepsilon^{\alpha\beta\gamma} S_m^\gamma, & m=n \\ S_n^\beta S_m^\alpha, & m\neq n \end{cases} \tag{2.60}$$

where α, β, γ = 1, 2, 3. The Hamiltonian and L -operator are [10]

$$H = -\frac{1}{2}\sum_n (S_n^1 S_{n+1}^1 + S_n^2 S_{n+1}^2 + \cosh\eta(S_n^3 S_{n+1}^3 - 1)), \tag{2.61}$$

$$L(n,u) = \begin{pmatrix} w_4 + w_3 S_n^3 & w S_n^- \\ w S_n^+ & w_4 - w_3 S_n^3 \end{pmatrix}, \quad \begin{aligned} & w_4 \pm w_3 = z^{\mp 1}, \\ & w = \tfrac{1}{2}(z^2 + z^{-2} - 2\cosh\eta)^{1/2}, \\ & z^2 = \frac{\sin u}{\sin(u+i\eta)}, \quad S^\pm = S^1 \pm i S^2. \end{aligned} \tag{2.62}$$

R -matrix coincides with (2.55) upon subsituting $u \to iu$, $v \to iv$, $2\gamma \to \eta$. Setting $z^2 = e^{-i\lambda\Delta}$, $\eta = \sqrt{\varkappa\Delta}$, $S_n^+/\sqrt{\Delta} \sim \psi^+(x), x = n\Delta$ one obtains in the limit $\Delta \to 0$ the Hamiltonian and the L -operator (2.3) for NS [10].

2. A lattice NS model by Ablowitz and Ladik [2]*) is formulated in terms of operators Ψ_n, Ψ_n^+ obeying the commutation relations

$$[\Psi_n, \Psi_m^+] = \hbar(1 + \Psi_n^+ \Psi_n)\delta_{nm} \tag{2.63}$$

The Hamiltonian, L -operator and R -matrix are

$$H = -\sum_n \Psi_n^+(\Psi_{n+1} - \Psi_{n-1}) + \frac{2\hbar}{\ln(1+\hbar)} \ln(1 + \Psi_n^+ \Psi_n) \tag{2.64}$$

$$L(n,z) = \begin{pmatrix} 1/z & -\Psi_n^+ \\ \Psi_n & z \end{pmatrix}, \quad z = \exp\varphi \tag{2.65}$$

$$R(\varphi) = \begin{pmatrix} a & & & \\ & b_+ & c & \\ & c & b_- & \\ & & & a \end{pmatrix} \qquad \begin{aligned} a &= \sinh(\varphi - \eta) \\ b_\pm &= e^{\pm\eta}\sinh\varphi \\ c &= -\sinh\eta \\ 2\eta &= \ln(1+\hbar) \end{aligned} \tag{2.66}$$

Letting $\Delta \to 0$ one obtains the corresponding quantities for NS.

3. A lattice NS model by Izergin and Korepin [28] deserves particular attention because it can provide a regular method of constructing the lattice completely integrable models from the continuous ones. The main idea of their approach could be expressed as follows. Let us look for the lattice L -operator in the form of a power series in Δ

$$L(x,\lambda;\Delta) = 1 + \Delta L^{(1)}(x,\lambda) + \Delta^2 L^{(2)}(x,\lambda) + \ldots \tag{2.67}$$

Putting $L^{(1)}(x,\lambda)$ to be equal to the continuous L -operator we may expect the higher corrections in L to be defined from (2.58) in classical case or (2.59) in quantum case. The r -matrix in (2.58) or R -matrix (2.59) is assumed to be the same as in the continuous case.

---

*) The Hamiltonian and quantum formulations of the model were proposed in [37, 38].

It turns out that for NS the series (2.67) can be put in a compact form [28]

$$L(n,\lambda) = \begin{pmatrix} -i\frac{\lambda}{2}\Delta + S_n^3 & i\varkappa S_n^+ \\ -iS_n^- & i\frac{\lambda}{2}\Delta + S_n^3 \end{pmatrix} \tag{2.68}$$

where $S_n^3 = 1 + \frac{\varkappa}{2}\Psi_n^+\Psi_n$, $S_n^- = \sqrt{1+\frac{\varkappa}{4}\Psi_n^+\Psi_n}\;\Psi_n$, (2.69)
$S_n^+ = (S_n^-)^*$, $[\Psi_n, \Psi_m^+] = \Delta\,\delta_{mn}$.
The formulae (2.69) are the same both in the classical and quantum cases.

The same program has been performed for SG [30, 31]. A disadvantage of the approach considered is that the lattice Hamiltonian turns out to be nonlocal in the quantum case. The fact is unimportant, however, if one uses the algebraic Bethe ansatz and is interested only in the continuous limit results.

2.5. General scheme of QSTM

To sum up let us list the main stages of QSTM. The list given below does not exhaust, of course, all technicalities of QSTM. It should be remarked also that the scheme presented can be modified drastically in applying to specific models.

0. CSTM for the classical model.

This preliminary stage includes finding the L -operator, Hamiltonian structure and r -matrix.

1. Quantum L -operator and R -matrix.

This stage includes choosing a lattice approximation (if necessary) for the classical and then for the quantum model (or immediately for the quantum one). The lattice L -operator may be found exactly or up to the leading order in Δ, as has been made in the first version of QSTM for SG [13].

2. Finding a pseudovacuum $|0\rangle$.

Before finding a pseudovacuum it is necessary to choose representation of the commutation relation for lattice field operators. The purpose of finding the pseudovacuum is to use some elements of the transition matrix $T(x_1, x_2; \lambda)$ as creation operators for pseudoparticles. The most popular condition for determining pseudovacuum is vanishing of some matrix elements of T (e.g., one of the off-diagonal elements for NS, SG, XYZ). In the simplest case (NS, XYZ) the pseudovacuum is a tensor product of single-site vectors [10]: $|0\rangle = \ldots \otimes e_i \otimes e_{i+1} \otimes \ldots$. For SG model $|0\rangle$ is a tensor product of two-site vectors [13, 20] . The most complicated structure of pseudovacuum occurs in the XYZ model [14] .

3. Writing the periodicity equations on the finite lattice (ring).

This stage includes choosing a generating functional of quantum integrals of motion (see Sec.5), solving the problem of extracting the Hamiltonian from the generating functional chosen (quantum trace formulae) and, finally, writing the periodicity equation.

The last problem is not sometimes so easy as in Sec.2.2 involving in case of L -operator of high matrix dimension tedious calculations with the commutation relation (2.36) [33-35, 41] .

4. Determining the physical vacuum Ω .

This stage includes an analysis of admissible solutions to the periodicity equations [42, 43] , removal of space cutoffs and, finally, writing the set of integral equations for the density of pseudoparticle momenta in the physical vacuum and excited states. It is preferable to remove the ultraviolet cutoff at the same time as the space cutoff instead of solving the system on the lattice and making $\Delta \to 0$ in the final results.

5. Calculation of physical characteristics: elementary excitations, their bound states and S -matrices.

See, for example, [13, 25, 26, 47] .

6. Inverse spectral transform and calculation of Green's functions.

Under the inverse spectral transform we understand here expressing the original field operators (e.g. Ψ , Ψ^+ for NS) in terms of the scattering data (e.g. A, A^+, B, B^+ for NS). In other words, it is the problem of finding the quantum generalization of the classical Gelfand-Levitan equation. Solution of this problem apparently gives rise to calculation of Green's functions.

At present the problem is solved only for NS in the repulsive case $\varkappa > 0$ [48, 52] . Some important improvements which allow one to consider the case $\varkappa < 0$ have been made recently in [44] .

Let us list now several topics closely connected with QSTM which have fallen beyond the scope of our review.

1. Thermodynamics of one-dimensional exactly soluble quantum systems.

Within the Bethe ansatz approach the problem was considered in the pioneering paper by C.N.Yang and C.P.Yang [67] for NS. Yang's approach has been then extended to XXX [68], XXZ and XYZ [69] and SG [70] models. Recently, within the QSTM approach to NS, H.B. Thacker et al. [57-59] has succeeded in rederiving Yang's results.

2. Models with isotopic symmetry.

Considering the models with many degrees of freedom due to a group of isotopic symmetry needs some complications in QSTM. Instead of a single algebraic Bethe ansatz there arises a hierarchy of higher Bethe ansatze [33-35, 41].

3. Direct methods of calculating the physical spectrum.

An interesting direction in QSTM originated in statistical physics is connected with attempts to avoid the stages 2-3 of the above list and to calculate the spectrum of the Hamiltonian immediately from the "first principles" such as properties of unitarity, analyticity and crossing symmetry [71-73].

4. Tetrahedron equations.

A new promising proposal by A.B.Zamolodchikov [7] can become a basis for a multidimensional generalization of QSTM.

2.6. List of exactly soluble models.

The list given below contains brief descriptions of the quantum models soluble by QSTM as well as some models which are solved by means of Bethe ansatz technique and other possible candidates for applying QSTM. We tried to make the list as complete as possible though we have excluded, on the one hand, the free fermion models (XY, impenetrable Bose gas, Ising-field-model [74, 75]) and, on the other hand, the models soluble via canonical transformation (massless Thirring model, Luttinger model, Federbusch model, Schwinger model etc. [75-76].

1. Nonlinear Schrödinger equation NS.

The model has been discussed already in subsection 2.1-2. It is to be added only that NS has a lot of generalisations.

The matrix generalisation of NS [77, 3] is defined in the classical case by the Hamiltonian

$$H=\int dx\, tr(\Psi_x^+\Psi_x+\varkappa\Psi^+\Psi\Psi^+\Psi) \tag{2.70}$$

where Ψ is the $n\times m$ matrix of dynamical variables $\Psi_{\alpha\beta}(x)$, $\Psi^+_{\gamma\delta}(x)$ being its hermitian conjugate. The Poisson brackets are standard

$$\{\psi_{\alpha\beta}(x), \psi^+_{\gamma\delta}(y)\} = i\delta_{\alpha\delta}\delta_{\beta\gamma}\delta(x-y)$$

The L-operator is (m+n) × (m+n) block-matrix

$$L(x,\lambda) = i\begin{pmatrix} -\frac{\lambda}{2} & \varkappa\Psi^+(x) \\ \Psi(x) & \frac{\lambda}{2} \end{pmatrix} \tag{2.71}$$

The quantisation of the matrix NS, like of the scalar one, resolves itself to normal ordering. The vector (m=1) NS has been considered in frame of Bethe ansatz method in [56-58, 78, 79] . Within QSTM the matrix NS has been treated in [33, 34] . A variant of the matrix NS including Fermi fields has been considered in [34] . The reduction problem for the matrix NS and a new completely integrable model, the $O(N)$ -invariant NS, are investigated in [40].

The lattice versions of the scalar NS [10, 28, 29, 37, 38] have been discussed in subsection 2.4. The inverse spectral transform for NS is proposed in [48, 52] . The thermodynamics of NS is discussed in [51, 67] .

2. XYZ-model and other lattice magnets.

The XYZ-model is formulated in terms of the spin- $\frac{1}{2}$ operators S^α_n (2.60). The XYZ Hamiltonian reads

$$H = -\frac{1}{2}\sum_n \left(J_1 S^1_n S^1_{n+1} + J_2 S^2_n S^2_{n+1} + J_3 S^3_n S^3_{n+1}\right) \tag{2.72}$$

The history of investigating XYZ-model and especially its degenerations, such as XXX ($J_1 = J_2 = J_3$), XXZ ($J_1 = J_2 \neq J_3$, see subsec 2.4) and XY ($J_3 = 0$) models is too long to touch it here. We must mention, however, the pioneering papers by R.J.Baxter [23--24] who solved the most general case $J_1 \neq J_2 \neq J_3$ of the XYZ model using a far reaching generalization of Bethe ansatz method which became one of the origins for QSTM. The paper [80] must also be mentioned which is devoted to the calculation of the spectrum of the XYZ Hamiltonian (2.72) within Baxter's method.

The QSTM was applied to XXZ-model in [10] and to the general XYZ-model in [14] . The L-operator for the XYZ-model is parametrized by the Jacobian elliptic functions and reads

$$L(n,u)=\sum_{\alpha=0}^{3} W_\alpha(u) S_n^\alpha \sigma_\alpha \tag{2.73}$$

where σ_α (α =1, 2, 3) are Pauli matrices, σ_0=1, and

$$W_0(u)=1, \qquad J_1 : J_2 : J_3 =$$
$$W_1(u)=sn(\gamma,k)/sn(u+\gamma,k), \qquad = 1 : dn(2\gamma,k) : cn(2\gamma;k),$$
$$W_2(u)=dn(u+\gamma,k)\,sn(\gamma,k)/sn(u+\gamma,k)\,dn(\gamma,k),$$
$$W_3(u)=cn(u+\gamma,k)\,sn(\gamma,k)/sn(u+\gamma,k)\,cn(\gamma,k).$$

The R -matrix has the same form as $L_n(u)$

$$R(u)=\sum_{\alpha=0}^{3} W_\alpha(u)\sigma_\alpha\otimes\sigma_\alpha=\begin{pmatrix} a & & & d \\ & b & c & \\ & c & b & \\ d & & & a \end{pmatrix} \tag{2.74}$$

$$a=W_0+W_3 \qquad c=W_1+W_2$$
$$b=W_0-W_3 \qquad d=W_1-W_2$$

The elliptic functions in (2.73-74) can degenerate into the trigonometric (hyperbolic) and the rational ones. The trigonometric degeneration (XXZ-model) is considered in subsec.2.4, the rational degeneration (XXX-model) reads

$$L_n(u)=1+\frac{\varkappa}{u+\varkappa}\sum_{\alpha=1}^{3} S_n^\alpha\sigma_\alpha \tag{2.75}$$

and has the obvious SU(2) symmetry.

The classical continuous analog of XYZ-model is the so called Landau-Lifshitz equation [60]

$$\vec{S}_t=\vec{S}\times\vec{S}_{xx}+\vec{S}\times J\vec{S}, \quad \vec{S}=(S^1,S^2,S^3), \quad |\vec{S}|=1,$$
$$J=diag(J_1,J_2,J_3), \quad J_1<J_2<J_3 \tag{2.76}$$

describing the spin waves in the ferromagnets. The Hamiltonian and the Poisson brackets for (2.76) are

$$H=\frac{1}{2}\int_{-\infty}^{\infty}dx\left[\vec{S}_x^2-(\vec{S},J\vec{S})+J_3\right] \tag{2.77}$$

$$\{S^{\alpha}(x), S^{\beta}(y)\} = \varepsilon^{\alpha\beta\gamma} S^{\gamma}(x)\delta(x-y) \tag{2.78}$$

The L-operator and r-matrix are

$$L(x,u) = -i\sum_{\alpha=1}^{3} r_{\alpha}(u) S^{\alpha}(x)\sigma_{\alpha}, \quad r(u) = \sum_{\alpha=1}^{3} r_{\alpha}(u)\sigma_{\alpha}\otimes\sigma_{\alpha} \tag{2.79}$$

where $r_{\alpha}(u) = \frac{\partial}{\partial\eta}\Big|_{\eta=0} w_{\alpha}(u,\eta)$.

It is to be noted that the XYZ-model can be considered as a lattice version of MT and SG models [81-82] . A wide class of lattice models generalizing XYZ-model is described in subsec.3.2 in connection with the so-called Yang-Baxter equation.

We proceed now to describe the relativistic completely integrable models. The first group of models (nn. 3-5) are the models including the scalar fields which can be considered as generalisations of the sine-Gordon model.

3. Sine-Gordon model (SG).

SG is the model of the selfinteracting relativistic scalar field φ taking its values on the circle $0 \le \varphi \le 2\pi$. The Lagrangian is

$$\mathcal{L} = \frac{1}{8\gamma}\left\{\frac{1}{2}\partial^{\mu}\varphi\,\partial_{\mu}\varphi - m^2(1-\cos\varphi)\right\} \tag{2.80}$$

The Hamiltonian structure is described in terms of the canonical variables $\varphi(x)$ and $\pi(x) = (8\gamma)^{-1}\partial_0\varphi(x)$.

$$\{\pi(x), \varphi(y)\} = \delta(x-y)$$

The Hamiltonian is

$$H = \int dx\left\{4\gamma\pi^2 + \frac{1}{8\gamma}\left(\frac{1}{2}\partial_x\varphi\,\partial_x\varphi + m^2(1-\cos\varphi)\right)\right\} \tag{2.81}$$

The equations of motion are

$$\dot{\varphi} = 8\gamma\pi\ , \quad \dot{\pi} = \frac{1}{8\gamma}\left(\varphi_{xx} - m^2\sin\varphi\right) \tag{2.82}$$

The classical L-operator for SG read

$$L=-i\begin{pmatrix} 2\gamma\pi & \frac{im}{4}\left(\lambda e^{-i\frac{\varphi}{2}}-\frac{1}{\lambda}e^{i\frac{\varphi}{2}}\right) \\ -\frac{im}{4}\left(\lambda e^{i\frac{\varphi}{2}}-\frac{1}{\lambda}e^{-i\frac{\varphi}{2}}\right) & -2\gamma\pi \end{pmatrix} \tag{2.83}$$

The r-matrix

$$r(\lambda/\mu)=\begin{pmatrix} 0&0&0&0\\ 0&b&c&0\\ 0&c&b&0\\ 0&0&0&0 \end{pmatrix} \quad b=\gamma\frac{\lambda^2+\mu^2}{\lambda^2-\mu^2},\ c=\frac{2\gamma\lambda\mu}{\mu^2-\lambda^2}$$

up to a change of variables is a classical variant (in the sense of (2.37)) of the XXZ-model's R-matrix (2.55).

Since the complete integrability of the classical SG had been established [1, 2] it became the object of the intensive study. It was studied by semiclassical methods [83] and within the bootstrap program [84] . The successful solution of SG within QSTM [13] remains one of the major triumphs of QSTM. A more recent treatment of SG on the base of a lattice approximation [30, 31] is presented in [20] .

The sinh-Gordon (ShG) equation is obtained from (2.82) by replacing $\sin\beta\varphi$ by $\sinh\beta\varphi$. The application of QSTM to ShG should not face any significant difficulties in comparison with SG. The explanation of the fact that it is not done yet may be that the expected spectrum is not very interesting (there are no solitons and bound states like in SG).

4. Super-symmetric sine-Gordon model (SSG).

The supersymmetric generalisation of SG has been solved via CSTM in [85]. The model describes the interaction of the scalar field and Majorana spinor field ψ . The Lagrangian is

$$\mathcal{L}=\frac{1}{2}\left[\partial^\mu\varphi\,\partial_\mu\varphi+i\bar{\psi}\gamma^\mu\partial_\mu\psi-\frac{m^2}{\beta^2}\sin^2\beta\varphi+im\bar{\psi}\psi\cos\beta\varphi\right] \tag{2.84}$$

The L-operator reads $(m=1,\ \beta=1)$

$$L=\frac{1}{2}\left\{\begin{pmatrix} i\lambda^2/2 & (\varphi_x+\varphi_t) & -\lambda\psi_1 \\ -(\varphi_x+\varphi_t) & -i\lambda^2/2 & 0 \\ \lambda\psi_1 & 0 & -i\lambda^2/2 \end{pmatrix}+\frac{i}{\lambda^2}\begin{pmatrix} \sin^2\varphi & \frac{1}{2}\sin 2\varphi & \lambda\psi_2\cos\varphi \\ \frac{1}{2}\sin 2\varphi & \cos^2\varphi & -\lambda\psi_2\cos\varphi \\ \lambda\psi_2\cos\varphi & -\lambda\psi_2\cos\varphi & 1 \end{pmatrix}\right. \tag{2.85}$$

The classical r-matrix up to a similarity transformation coincides with that of the Fermi MT model.

The only exact quantum result known for SSG is the calculation of the S-matrix within the bootstrap program [86].

5. Relativistic field models connected with root systems (RS).

Let $\mathcal{R}$ be a root system, $\mathcal{A}$ the set of admissible roots, $\{H_k, E_\alpha\}$ Cartan-Weyl basis, n the Coxeter number, $\rho(\alpha)$ the height of the root α (mod n), l the range of $\mathcal{R}$ [87] . Then the system of l relativistic fields φ_k defined by the Lagrangian

$$\mathcal{L}=\sum_{k=1}^{l}\partial_\mu\varphi^k\partial^\mu\varphi^k-\frac{1}{2}\sum_{\alpha\in\mathcal{A}}\exp 2\varphi_\alpha;\quad \varphi_\alpha=\sum_{k=1}^{l}\varphi^k\alpha_k \tag{2.86}$$

and the equations of motion

$$\varphi^k_{tt}-\varphi^k_{xx}=\frac{1}{2}\sum_{\alpha\in\mathcal{A}}\alpha_k\exp 2\varphi_\alpha \tag{2.87}$$

turns out to be completely integrable [87] . The corresponding L-operator is

$$L(x,\lambda)=\sum_{k=1}^{l}\pi^k(x)H_k+\sum_{\alpha\in\mathcal{A}}e^{\varphi_\alpha}(\lambda E_\alpha+\frac{1}{\lambda}E_\alpha);\quad \pi^k=\partial_0\varphi^k \tag{2.88}$$

The corresponding classical r-matrix is presented in Sec.4, formula (4.2).

In case $\mathcal{R}$ belongs to the type A_{l-1} the formulae (2.86-2.88) take the form

$$\mathcal{L}=\sum_{k=1}^{l}\partial^\mu\varphi^k\partial_\mu\varphi^k-\frac{1}{2}\sum_{k=1}^{l}\exp 2(\varphi^k-\varphi^{k+1}) \tag{2.89}$$

$$\varphi^k_{tt}-\varphi^k_{xx}=\frac{1}{2}\{\exp 2(\varphi^{k-1}-\varphi^k)-\exp 2(\varphi^k-\varphi^{k+1})\} \tag{2.90}$$

$$L(x,\lambda)=\sum_{k=1}^{l}\pi^k(x)e_{kk}-\frac{1}{2}\sum_{k=1}^{l}e^{2(\varphi^k-\varphi^{k+1})}(\lambda e_{k,k+1}+\frac{1}{\lambda}e_{k,k-1}) \tag{2.91}$$

$$(e_{ij})_{ab}=\delta_{ia}\delta_{jb}$$

Imposing some constraints on the field φ^k one can obtain many other completely integrable systems [87] . For example, the reduction (l=3) $\varphi^1=-\varphi^3=\varphi$, $\varphi^2=0$ leads to the equation

$$\varphi_{tt} - \varphi_{xx} = e^{2\varphi} - e^{-4\varphi} \tag{2.92}$$

The application of QSTM to the models described is limited at present to the stage 1 of the list given in Sec.2.5 [19, 27, 38, 55] .

The next two models describe the interacting massive spinor fields.

6. Massive Thirring model (MT).

See Sec.2.3.

7. Bukhvostov-Lipatov model (BL).

The BL Lagrangian [88] reads

$$\mathcal{L} = \sum_{a=1}^{2} \bar{\psi}_a (i\gamma^\mu \partial_\mu - m)\psi_a - g j_1^\mu j_{2\mu} \; ; \; j_a^\mu = \bar{\psi}_a \gamma^\mu \psi_a \, . \tag{2.93}$$

The model is studied in [88] within the Bethe ansatz method. Neither CSTM nor QSTM investigation of the model has been undertaken yet.

Of special interest are the models (n.n.8-10) which are massless in the classical case but being quantized reveal a massive spectrum due to dimensional transmutation. None of the models has been solved yet by QSTM.

8. Isotopic massless Thirring model (IMT).

The model is defined by the $SU(N)$ -symmetric Lagrangian

$$\mathcal{L} = \sum_{a=1}^{N} i\bar{\psi}_a \gamma^\mu \partial_\mu \psi_a - \sum_{a,b,c,d} \bar{\psi}_a \gamma^\mu \psi_c \, V_{ab;cd} \, \bar{\psi}_b \gamma_\mu \psi_d \; ; \; V = g_0 I + g P \tag{2.94}$$

which in case $g_0 = 0$ coincides with the Lagrangian of the chiral Gross-Neveu model

$$\mathcal{L} = \sum_{a=1}^{N} i\bar{\psi}_a \gamma^\mu \partial_\mu \psi_a + g\left((\bar{\psi}\psi)^2 - (\bar{\psi}\gamma_5\psi)^2 \right) . \tag{2.95}$$

The model (2.95) has been proposed in [89] and studied semiclassically in [90] . S-matrix of the model has been discussed in frame of the bootstrap program in [91] . Recently the model has been solved via Bethe ansatz method starting from the Lagrangian (2.94) in [92, 93, 35] and from (2.95) in [94] . It is to be noticed that the Lagrangians (2.94) and (2.95) turn out to yield the same final result for the physical spectrum, since the S-matrices of pseudoparticles over the pseudovacuum $|0\rangle$ coincide for (2.94) and (2.95).

An interesting generalization of the model has been proposed recently by V.N.Dutyshev [95] in case N=2. He has shown that the model defined by

$$\mathcal{L} = i\bar{\psi}\gamma^\mu\partial_\mu\psi - \sum_{a=1}^{3} g_a(\bar{\psi}\gamma_\mu\sigma^a\psi)^2 \tag{2.96}$$

is also soluble via Bethe ansatz technique. A further generalization of (2.94) is considered in Sec.3.2.

9. The Gross-Neveu model (GN).

The model is described by the O(N)-symmetric Lagrangian

$$\mathcal{L} = i\bar{\varphi}\gamma^\mu\partial_\mu\varphi - g(\bar{\varphi}\varphi)^2 \tag{2.97}$$

where φ^a, $a = 1, 2, \ldots, N$ are real Majorana spinors. In case $N = 2M$ there is an equivalent Lagrangian

$$\mathcal{L} = i\bar{\psi}\gamma^\mu\partial_\mu\psi - g(\bar{\psi}\psi)^2$$

where $\psi^a = \varphi^a + i\varphi^{a+M}$ $(a = 1, \ldots, M)$. The model has been proposed in [89] and studied in the semiclassical approximation in [96] The known exact results are: CSTM [97, 98] and S-matrix calculation within the bootstrap program [84, 99].

10. Chiral fields (CF).

The principal chiral field [100] is the field g taking its values in a Lie group G described by the Lagrangian

$$\mathcal{L} = \mathrm{tr}(g^{-1}\partial_\mu g)(g^{-1}\partial^\mu g), \quad g \in G. \tag{2.98}$$

The equation of motion are

$$\partial^2 g = (\partial_\mu g)g^{-1}(\partial^\mu g). \tag{2.99}$$

There is a lot of reductions of (2.99) corresponding to various symmetric spaces. The simplest of them is the nonlinear σ-model which is described in terms of the field $n=(n_1, n_2, n_3)$ taking its values on the sphere S^2, $\vec{n}^2=1$. The Lagrangian and the equations of motion are

$$\mathcal{L}=(\partial_\mu \vec{n}, \partial^\mu \vec{n}), \qquad \partial^2 \vec{n} + \vec{n}(\partial_\mu \vec{n}, \partial^\mu \vec{n})=0. \tag{2.100}$$

The Hamiltonian structure is described in terms of the variables $\vec{n}(x)$ and $\vec{l}(x)=\vec{n}\times\vec{n}_t$ and is defined by the Poisson brackets

$$\begin{aligned} &\{ n_a(x), n_b(y)\} = 0, \\ &\{ l_a(x), n_b(y)\} = \varepsilon^{abc} n_c(x)\,\delta(x-y), \\ &\{ l_a(x), l_b(y)\} = \varepsilon^{abc} l_c(x)\,\delta(x-y) \end{aligned} \tag{2.101}$$

and the Hamiltonian

$$H=\frac{1}{2}\int dx\,(\vec{n}_x^{\,2} + \vec{l}^{\,2}). \tag{2.102}$$

I.V.Cherednik has shown recently [101] that introducing an anisotropy in the Lagrangian (2.100)

$$\mathcal{L}=(\vec{J}_\mu, V\vec{J}^\mu), \quad \vec{J}_\mu=\vec{n}\times\partial_\mu\vec{n}, \quad V=\mathrm{diag}(V_1,V_2,V_3) \tag{2.103}$$

leaves the model completely integrable. The result is quite similar to that of Dutyshev for IMT model (2.96) and can also be generalized to arbitrary solutions of the Yang-Baxter equation (see Sec.3.2).

In the current literature there is a lot of results for chiral fields concerning their classical complete integrability [102] , quantum integrals of motion [103, 104] S-matrices [84, 86] .

Though the classical variants of the IMT, GN and CF models have been solved via CSTM the L-operators proposed in [97, 98, 102] do not contain the angle variables being thus of no use for QSTM. The problem of applying QSTM to the models remains still unsolved.

We finish our list with two models for which the attempts to apply QSTM still meet with failure.

11. Hubbard model (HM).

The model is formulated in terms of two fermionic fields ψ_n^α (α =1, 2) on the lattice. The SU(2)-invariant Hamiltonian [105] reads

$$H=\sum_n (\psi_n^{+\alpha}\psi_{n+1}^{\alpha}+\psi_{n+1}^{+\alpha}\psi_n^{\alpha}+g\psi_n^{+1}\psi_n^{+2}\psi_n^{2}\psi_n^{1}) . \qquad (2.104)$$

The energy spectrum of H has been determined in [105, 106] by means of Bethe ansatz technique. The model can be considered as a lattice approximation for the IMT model [107] and for the fermionic vector NS model (depending on the vacuum Ω chosen). A generalization of the model including 4 fermionic field has been considered in [108].

12. Toda chain (TC).

The model represents a chain of point particles described by coordinates q_n and momenta p_n with canonical Poisson brackets $\{p_m, q_n\} = \delta_{mn}$ (in the classical case) or commutation relations $[p_m, q_n] = -i\delta_{mn}$ (in the quantum case). The Hamiltonian is

$$H=\sum_n \left(\frac{1}{2}p_n^2 + \exp(q_{n+1}-q_n)\right) \qquad (2.105)$$

and L-operator

$$L_n(\lambda) = \begin{pmatrix} \lambda - p_n & e^{q_n} \\ e^{-q_n} & 0 \end{pmatrix} \qquad (2.106)$$

are the same both in the classical and quantum cases. The r(R)-matrix is the same as in case of NS (2.12, 37) [15, 32] .For the CSTM of the model see [1] . The known quantum results concern mainly the integrals of motion [109, 110] . For generalizations of (2.105) connected with root systems see [111] . Another possible generalization is the matrix $GL(N)$ -invariant Toda chain [112] which is formulated in terms of variables g taking values in the group GL(N). The Lagrangian is [32]

$$\mathcal{L}=\sum_n \mathrm{tr}\left(\frac{1}{2}A_n^2-B_n\right), \quad A_n=(\partial_t g_n)g_n^{-1}, \quad B_n=g_{n+1}g_n^{-1} . \qquad (2.107)$$

The model (2.107) was considered in [112] as a possible lattice approximation of the principal chiral field.

We finish this Section with a table representing the present state (spring, 1981) of the models listed above. The rows correspond to the models taken in the same order as in the above list, the columns correspond to the stages 1-6 of the QSTM listed in Sec.

2.5. The two extra columns CSTM and BA denote the classical spectral transform method and coordinate Bethe ansatz respectively. In the squares the year of the first publication and some references are given. A blank square means that the corresponding problem for the corresponding model has not yet been solved.

Table 1.

	CSTM	BA	1	2	3	4	5	6
1. NS	1971 [1-3, 33,77]	1963 [56-58, 78,79]	1979 [9,15, 18]	no problem	1963 [58,15]	1963 [58]	1963 [58]	1979 [44, 48,52]
2. XYZ	1979 [60]	1973 [24,14]	1972 [23,24, 14]	1973 [24,14]	1972 [23,24, 14]	1972 [23,24, 14]	1973 [80]	
3. SG	1974 [1,3]		1979 [13,20]	1979 [13,20]	1979 [13,20]	1979 [13,20]	1979 [13,20]	
4. SSG	1978 [85]							
5. RS	1980 [87]		1980 [19,27, 55]					
6. MT	1977 [62,63]	1965 [64]	1980 [19]	no problem	1979 [47,25]	1979 [47,25, 26]	1979 [47,25, 26]	
7. BL		1980 [88]		no problem	1980 [88]	1980 [88]	1980 [88]	
8. IMT	1978 [97,98]	1979 [92-95]		no problem	1979 [92-95]	1979 [92-94]	1979 [92-94]	
9. GN	1978 [97,98]							
10. CF	1975 [102]							
11. HM		1968 [105]		no problem	1968 [105]	1968 [105]	1968 [106]	
12. TC	1974 [1-3]		1979 [15,32]					

3. YANG-BAXTER EQUATION AND ITS APPLICATIONS

In the previous Section we have presented a general sketch of QSTM. Now we shall concentrate on a couple of specific topics which are closer to us.

The first topic which is the subject of the present and the next sections is the Yang-Baxter equation (known also as "factorization equation", "star-triangle" relation). The Yang-Baxter equation (YBE)

$$\begin{aligned} {}_{\alpha\alpha'}R_{\gamma\gamma'}(u-v)\ {}_{\gamma\alpha''}R(u)_{\beta\gamma''}\ {}_{\gamma'\gamma''}R_{\beta'\beta''}(v) = \\ = {}_{\alpha'\alpha''}R_{\gamma'\gamma''}(v)\ {}_{\alpha\gamma''}R_{\gamma\beta''}(u)\ {}_{\gamma\gamma'}R_{\beta\beta'}(u-v) \end{aligned} \tag{3.1}$$

is a functional equation for a four-indices function ${}_{\alpha\beta}R_{\gamma\delta}(u)$ of a parameter u which we shall call the spectral parameter. The indices are assumed to run from 1 to N. Over the repeated indices the summation is assumed.

The YBE was introduced in [23, 78] and has numerous applications in the theory of completely integrable quantum and classical systems and of the exactly soluble models of statistical physics. Before going to the applications of YBE let us introduce some useful notation.

The four-indices quantity ${}_{\alpha\beta}R_{\gamma\delta}$ can obviously be interpreted as an operator R in the tensor product space $C^N \otimes C^N$. In the space $C^N \otimes C^N \otimes C^N$ we introduce three operators R_{12}, R_{13}, R_{23} corresponding to the three canonical embeddings of $C^N \otimes C^N$ into $C^N \otimes C^N \otimes C^N$ (for example, $R_{12} = R \otimes I_N$, $R_{23} = I_N \otimes R$). Thus, (3.1) can be rewritten in the operator form

$$R_{12}(u-v)\, R_{13}(u)\, R_{23}(v) = R_{23}(v)\, R_{13}(u)\, R_{12}(u-v). \tag{3.2}$$

A solution R to YBE we shall call Yang-Baxter bundle (YBB).

We must warn the reader that in some papers on QSTM [13-15,20] another operator $\check{R}$ is used which differs from R introduced above by the permutation operator $\mathcal{P}$

$$\check{R} = \mathcal{P} R, \tag{3.3}$$

$${}_{\alpha\beta}\mathcal{P}_{\gamma\delta} = \delta_{\alpha\delta}\, \delta_{\beta\gamma}. \tag{3.4}$$

The YBE in terms of $\check{R}$ reads [27]

$$(I\otimes \check{R}(u-v))(\check{R}(u)\otimes I)(I\otimes \check{R}(v))= =(\check{R}(v)\otimes I)(I\otimes \check{R}(u))(\check{R}(u-v)\otimes I). \qquad (3.5)$$

We shall consider also the classical analogue of the YBE

$$[r_{12}(u-v), r_{13}(u)] + [r_{12}(u-v), r_{23}(v)] + [r_{13}(u), r_{23}(v)] = 0 \qquad (3.6)$$

which can be obtained by inserting the semiclassical expansion (2.37) for R into (3.2) and taking the terms of order $\hbar^2$.

3.1. Origin of YBE.

The YBE has been introduced in a paper by C.N.Yang [78]. His result, slightly modernized, can be expressed as follows. Consider the nonrelativistic quantum theory of N Bose fields described by the Hamiltonian

$$H = \int dx \left(\psi^{+}_{\alpha x} \psi_{\alpha x} + \psi^{+}_{\alpha} \psi_{\beta}\, {}_{\alpha\beta}V_{\gamma\delta}\, \psi_{\gamma} \psi_{\delta} \right). \qquad (3.7)$$

The two-particle S-matrix $S(k_1 - k_2)$ considered as an operator in $C^N \otimes C^N$ is simply the Cayley transformation of the potential

$$S(k) = (k + iV)^{-1}(k - iV). \qquad (3.8)$$

Yang's result is that if S(k) satisfies the YBE (3.2) then the exact eigenfunctions of H can be constructed by means of Bethe ansatz method. Strictly speaking, Yang considered a very specific potential $V = \frac{1}{2}(I + \mathcal{P})$ but his reasoning can without any change be applied to the general case.

The YBE (3.2) applied to the S-matrix (3.8) expresses the property of factorization of three-particle S-matrix. It means that the three-particle S-matrix can be represented as the product of three two-particle ones and the result of three-particle scattering does not depend on the order of two-particle collisions. In recent years it has been recognized(see e.g. [84]) that factorization of S-matrix is a manifestation of complete integrability of a model. For relativistic completely integrable models, the use of factorization, unitarity and crossing symmetry of S-matrix gives an opportunity to calculate S-matrix up to the well-known CDD ambiguity [84]. Encouraging is the fact that the S-matrices found for SG amd MT models [84] within the dynamical approach coincide with those found previously within the bootstrap program.

Returning to Yang's paper [78] it is to be noted that the problem of describing all the potentials V in (3.7) which generate factorized S-matrices is not solved yet. A particular solution, in addition to Yang's one, is the O(N)-invariant potential [40]

$$ {}_{\alpha\beta}V_{\gamma\delta} = \frac{\varkappa}{2}\left(\delta_{\alpha\gamma}\delta_{\beta\delta} + \delta_{\alpha\delta}\delta_{\beta\gamma} - \delta_{\alpha\beta}\delta_{\gamma\delta}\right). \tag{3.9} $$

It is natural to state the same problem also for differential operators of higher order and matrix differential operators, for example, for Dirac operator. In papers [95] and [88] examples are given of the potentials yielding factorized S-matrices for massless and massive Dirac operators respectively. Of certain interest is also the inverse problem: to find a differential operator and a δ-function potential generating a given factorized S-matrix.

Independently the YBE has arisen in R.J.Baxter's paper [23] as the commutativity condition for the transfer matrices of the so-called eight-vertex model in the lattice statistics. In this context YBE generalizes the famous star-triangle relation introduced by L.Onsager [22] for the Ising model.

Within QSTM the YBE arises as the condition on R-matrix which follows under certain assumptions from the fundamental equation (2.36). Using indices one can write (2.36) as

$$ {}_{\alpha\alpha'}R_{\beta\beta'}(u-v)\,T_{\beta\gamma}(u)\,T_{\beta'\gamma'}(v) = T_{\alpha'\beta'}(v)\,T_{\alpha\beta}(u)\,{}_{\beta\beta'}R_{\gamma\gamma'}(u-v). \tag{3.10} $$

Introducing the notation $T^{(1)} = T\otimes I_N \otimes I_N$, $T^{(2)} = I_N \otimes T\otimes I_N$, $T^{(3)} = I_N\otimes I_N \otimes T$ we get $(m\neq n)$

$$ T^{(m)}(u)\,T^{(n)}(v) = R^{-1}_{mn}(u-v)\,T^{(n)}(v)\,T^{(m)}(u)\,R_{mn}(u-v). \tag{3.11} $$

Consider now the product $T^{(1)}(u_1)T^{(2)}(u_2)T^{(3)}(u_3)$ and apply thrice (3.11). This can be made in two ways: either along the scheme $(123)\to(213)\to(231)\to(321)$ or $(123)\to(132)\to(312)\to(321)$. The result must be the same:

$$R_{23}^{-1}(u_2-u_3)R_{13}^{-1}(u_1-u_3)R_{12}^{-1}(u_1-u_2)T^{(3)}(u_3)T^{(2)}(u_2)T^{(1)}(u_1)R_{12}\times$$
$$\times R_{13}R_{23} = R_{12}^{-1}R_{13}^{-1}R_{23}^{-1}T^{(3)}(u_3)T^{(2)}(u_2)T^{(1)}(u_1)R_{23}R_{13}R_{12}. \tag{3.12}$$

It follows from (3.12) that the c -number matrix $\mathcal{R} = R_{12}R_{13}R_{23}R_{12}^{-1}R_{13}^{-1}R_{23}^{-1}$ commutes with the q -number matrix $\mathcal{T} = T^{(3)}(u_3)T^{(2)}(u_2)T^{(1)}(u_1)$. Taking matrix elements $\langle a|\mathcal{T}|b\rangle$ between arbitrary quantum states $|a\rangle$, $|b\rangle$ we obtain a variety of c -number matrices commuting with $\mathcal{R}$. It is natural to assume that the set $\langle a|\mathcal{T}|b\rangle$ is rich enough for its commutant to be trivial. If the assumption is true $\mathcal{R}$ must be scalar matrix. Noting then that $\det\mathcal{R}=1$ we arrive at

$$\mathcal{R} = I \tag{3.13}$$

which is equivalent to the YBE (3.2).

The above reasoning is not rigorous, of course, and, even more, there is a counterexample [40]. Nevertheless, as a matter of fact, for the majority of solved models the R-matrix satisfies YBE.

3.2. Applications of YBE

In this Subsection we show how from given YBB a set of completely integrable models can be constructed.

A YBB is called regular if it satisfies the condition

$$R(u)\big|_{u=0} = \mathcal{P}. \tag{3.14}$$

We shall show below that each regular YBB can be considered as L-operator for some exactly soluble quantum chain. Our reasoning is very much the same as Baxter's one [24] in connection with the XYZ model.

The quantum system in question is a closed ring of M "atoms" each of which has N quantum states. Thus, the space of quantum states is $\mathcal{H}_q = V_1 \otimes V_2 \otimes \ldots \otimes V_M$ ($V_1 = \ldots = V_M = C^N$). The L-operator $L_n(u)$ is considered as $N \times N$ matrix whose elements are operators in V_n. With the use of indices $L_n(u)$ is defined by

$$_{\alpha\alpha'}(L_n(u))_{\beta\beta'} = {}_{\alpha\alpha'}R_{\beta\beta'}(u) \tag{3.15}$$

the indices α, β being the matrix ones and α', β' being the quantum ones, $R(u)$ being a given regular YBB. It is easy to see that

YBE (3.2) implies (2.59) with the same R-matrix as in (3.15). The transition matrix T(1, M; u) is defined by (2.57). Due to (2.40), its matrix trace t(u)= tr T (1, M; u) is a commuting one-parametric family of operators. The last thing to do with t(u) is to extract a local Hamiltonian from it. M.Lüscher has shown [82] that $\ln(t(0)^{-1}\, t(u))$ is the generating function for commuting local quantities

$$\mathcal{J}_n = \frac{d^n}{du^n} \ln\left(t(0)^{-1}\, t(u)\right)\Big|_{u=0} . \tag{3.16}$$

Locality means here that $\mathcal{J}_n$ is a sum of operators each of which acts on no more than $n+1$ adjacent sites. In particular, $\mathcal{J}_1$ is expressed in terms of two-point density $H_{n,n+1}$

$$\mathcal{J}_1 = \sum_{n=1}^{M-1} H_{n,n+1} + H_{M,1} \tag{3.17}$$

where

$$H_{n,n+1} = \left(\frac{d}{du} R_n \quad (u)\Big|_{u=0}\right) \mathcal{P}_{n,n+1} . \tag{3.18}$$

Though Lüscher considered the XYZ model, his proof uses only the regularity property of R and thus can be applied in the general case.

A special comment is needed concerning the problem of completeness of the set t(u) of integrals of motion. Completeness means here that every operator commuting with $\mathcal{J}_1$ (3.17-18) is a function of t(u). A rigorous proof of completeness is available only in case of XXX model [113] . Nevertheless, the completeness of t(u) seems to be highly probable also for other models in the case of N=2, e.g. for XYZ model. In the case N>2, however, it is seen from CSTM, that the set t(u) might be incomplete. The problem of extra integrals of motion is discussed in Section 5.

A cherished dream of everybody who deals with the completely integrable systems is find an algorithm for judging if some given nonlinear evolution equation is or isn't completely integrable. Though the existence of the universal algorithm seems to be hardly probable, for some special classes of equations there might be such criteria of complete integrability. For example there is a simple condition for a quantum Hamiltonian with two-point density to be obtained from a regular YBB according to (3.15-18).

The idea of deriving the criterion is to expand YBE (3.2) in powers of u and V by inserting

$$R(u) = (I + H^{(1)}u + H^{(2)}u^2 + \dots)\mathcal{P} \qquad (3.19)$$

into (3.2). According to (3.18) $H^{(1)}$ is the two-point Hamiltonian density. It turns out that $H^{(2)}$ is defined from YBE (3.2) up to a scalar term

$$H^{(2)} = \frac{1}{2}(H^{(1)})^2 + \text{const}$$

(This follows from the fact that R(u) can always be multiplied by a scalar factor). The first nontrivial condition on $H^{(1)}$ is obtained as the solubility condition for linear equations defining $H^{(3)}$. The most elegant form for the condition proposed by N.Yu.Reshetikhin [114] reads as follows . It is necessary that the double commutator

$$[H^{(1)}_{12} + H^{(1)}_{23}, [H^{(1)}_{12}, H^{(1)}_{23}]] = X_{12} - X_{23} \qquad (3.20)$$

could be represented as the difference of some two-point quantities.

The condition (3.20) is effectively a set of homogenious cubic equations on $H^{(1)}$. The question if (3.20) is also a sufficient condition for restoring the whole YBB from H remains still unsolved. It is to be noted that for the Hubbard's model Hamiltonian (2.104) the condition (3.20) is not valid [114] . It means that there exist some schemes of complete integrability of lattice Hamiltonians which differ from the one discussed (3.15-18).

The quantum chain is not the only completely integrable system corresponding to the given YBB.

Consider the model of N interacting massless spinor fields $\psi_a = \begin{pmatrix} \psi_{1a} \\ \psi_{2a} \end{pmatrix}$ described by the Lagrangian

$$\mathcal{L} = \sum_{a=1}^{N} i\bar{\psi}_a \gamma^\mu \partial_\mu \psi_a - \sum_{a,b,c,d} \bar{\psi}_a \gamma^\mu \psi_c \, V_{ab;cd} \, \bar{\psi}_b \gamma_\mu \psi_d , \qquad (3.21)$$

$$\mathcal{P}V = V\mathcal{P} .$$

The corresponding Hamiltonian is

$$H = \int dx \, (-i\psi_a^+ \gamma^5 \partial_x \psi_a + 4\psi_{1a}^+ \psi_{2b}^+ \, {}_{ab}V_{cd} \, \psi_{2d} \psi_{1c}) . \qquad (3.22)$$

The S-matrix of the pseudoparticles over the pseudovacuum $|0\rangle$ defined by $\psi_{ia}|0\rangle=0$ is simply the Cayley transformation of the potential V

$$S(\sigma)=(\sigma+iV)^{-1}(\sigma-iV) \tag{3.23}$$

where $\sigma=\pm 1$ is the chirality. Since the S-matrix (3.23) does not depend on the momentum it is the function on the finite set $\{-1, 0, 1\}$. It is easy to see that we can satisfy the factorization equation (3.2) for S by choosing V as the inverse Cayley transformation $iV=\sigma(1+S(\sigma))^{-1}(1-S(\sigma))$ of some regular Yang-Baxter bundle R taken at an arbitrary value u_0 of its spectral parameter and satisfying the condition $R(-u_0)=R^{+}(u_0)=R^{-1}(u_0)$. A more detailed discussion of the model (3.21) will be published elsewhere. The case of R being Baxter's bundle (2.74) has been considered by V.N.Dutyshev [95] .

Let us turn now to the applications of the classical Yang-Baxter equation (3.6). The classical results are quite similar to the quantum ones. As in the quantum case, given a classical YBB one can construct a completely integrable model of ferromagnetic type [60, 115] . For example, the model corresponding to the classical analog of Baxter's XYZ solution (2.79) is the Landau-Lifshitz equation (2.76).

On the other hand, like in the quantum case, the classical YBB generates some classical massless relativistic models, namely the classical analog of the generalized isotopic massless Thirring model (3.21) and a generalized chiral field (for the N=2 case see [101]). The detailed description of the models mentioned will be published elsewhere.

To finish this Section let us list some recent advances of the YBE theory.

At present, a lot of particular solutions to YBE has been found (an extensive list is contained in the review [19]). At the same time, much progress has been made in the general YBE theory. I.V. Cherednik [116] and A.B.Belavin [115] have proposed powerful and rather general methods for solving YBE. I.M.Krichever [117] has classified YBB, for N=2 in the generic case. A.B.Zamolodchikov [7] has proposed a promising multidimensional generalisation of YBE. A group-theoretical approach to YBE has been proposed in [39] .

We have already mentioned that the list of particular solutions to YBE grows extensively. In this connection, of great interest is the problem of classifying them and isolating some regular series of YBBs. A method of generating infinite series of YBBs based on group-theoretical considerations is discussed in the next Section.

4. AN ALGEBRAIC APPROACH TO YBE

The idea of the algebraic approach to YBE which is developed in the present Section can be most easily explained by considering at first the classical YBE (3.6). Because the equation (3.6) is written in terms of commutators only, it may be regarded as an abstract Lie algebra equation. Given a solution r of (3.6) in a certain Lie algebra $\mathcal{G}$, i.e. $r \in \mathcal{G} \otimes \mathcal{G}$ we can obtain many particular solutions to (3.6) by considering particular linear representations of $\mathcal{G}$. The following examples illustrate the point.

1. Let $\{e_i\}$ be a basis of a semisimple Lie algebra $\mathcal{G}$, k its Killing form. Put $k_{ij} = k(e_i, e_j)$, $k^{ij} k_{jl} = \delta^i_l$. Then r(u)

$$r(u) = \frac{1}{u} \sum_{i,j} k^{ij} e_i \otimes e_j \tag{4.1}$$

satisfies (3.6). Moreover, A.B.Belavin has recently claimed [155] that every solution to the classical YBE of the form r/u is given by (4.1).

2. Let $\mathcal{R}$ be a root system, $\{H_j, E_\alpha\}$ Cartan-Weyl basis, n the Coxeter number, $p(\alpha)$ the height of the root α (mod n). Then r(u) given by

$$r(u) = \operatorname{cotanh} \frac{n}{2} u \sum_j H_j \otimes H_j + \sum_{\alpha \in \mathcal{R}} \frac{\exp(u(\frac{n}{2} - p(\alpha)))}{\sinh \frac{n}{2} u} E_\alpha \otimes E_{-\alpha} \tag{4.2}$$

satisfies (3.6). The solution (4.2) serves as the r-matrix for the two-dimensional Toda lattice model (2.86).

3. There exist also completely integrable models corresponding to the nonreduced root systems, e.g. BC_n [87]. The corresponding r-matrices can be calculated and be shown to satisfy YBE (3.6). We do not present here the corresponding cumbersome formulae (for BC_1 case see [27]).

On substituting the expansion

$$r(u) = \frac{r_{-1}}{u} + r_0 + r_1 u + \dots \tag{4.3}$$

into YBE (3.6) it is easy to see that the leading term r_{-1}/u must also satisfy (3.6). This fact combined with Belavin's result [115] mentioned above can provide a basis for classification of the classical YBB, according to the corresponding Lie algebrae. It is natural to assume that every classical YBE generates as many quantum ones as there are representations corresponding to its Lie algebra $\mathcal{G}$. The basis for such assumption is provided by recently found YBBs [118, 119] which can be interpreted as corresponding to the vector representations of SO(3) and SO(4). Belavin's solution [115] corresponds to the fundamental representation of SU(N).

Below we describe a construction of YBBs for higher finite dimensional representations of SU(2) from the fundamental one. All the proofs will be given for Baxter's XYZ-bundle (2.74) which corresponds in the above sense to the spin - 1/2 representation of SU(2). Some generalizations to SU(N) case can be found in [39].

4.1. The construction of YBBs

The construction described below is based on the fact that the spin-1-representation of SU(2) can be obtained as the symmetrized tensor product of 2 spin-1/2 representations.

It is instructive to consider Baxter's bundle R_{12} (2.74) as an S-matrix describing the scattering of two spin-1/2 particles (labelled by 1 and 2). Consider (as an absolutely formal object) a composite particle 12 consisting of two particles 1 and 2 with the rapidities $u+u_0$ and $u-u_0$ respectively. The scattering of the composite particle by the third particle with the rapidity v is described by the S-matrix $R_{12,3}$

$$R_{12,3}(u-v) = R_{13}(u-v+u_0)\,R_{23}(u-v-u_0). \qquad (4.4)$$

The space of states of the composite particle 12 is fourdimensional and is decomposed into one-dimensional (spin-o) and three-dimensional (spin -1) subspaces. However, in the course of scattering by the particle 3 these states are, generally speaking, mixed. The state of spin — 1 will not be destroyed in the course of scattering if the triangularity condition

$$P_{12}^{-}\,R_{12,3}\,P_{12}^{+} = 0 \qquad (4.5)$$

is satisfied, where P_{12}^{+} is the projector onto the spin -1 subspace (symmetrizer) and P_{12}^{-} is the projector onto the spin -o subspace (antisymmetrizer).

For Baxter's bundle R (2.74) the condition (4.5) is satisfied at $u_0 = -\gamma$ due to the fact that at the special value of the spectral parameter $u=-2\gamma$ Baxter's bundle (2.74) turns (up to an insignificant multiplier) into the antisymmetrizer P^-_{12}

$$-\frac{sn(u+\gamma;k)}{4sn(\gamma;k)} R_{12}(u;k)\Big|_{u=-2\gamma} = P^-_{12} . \tag{4.6}$$

On substituting $u = u - \gamma$, $v = u + \gamma$ into YBE (3.2) we obtain ($2\gamma \equiv \eta$)

$$P^-_{12} R_{13}(u-\tfrac{\eta}{2}) R_{23}(u+\tfrac{\eta}{2}) = R_{23}(u+\tfrac{\eta}{2}) R_{13}(u-\tfrac{\eta}{2}) P^-_{12} . \tag{4.7}$$

On putting $u_0 = -\frac{\eta}{2}$ in (4.4) and using (4.7) and $P^-_{12} P^+_{12} = 0$ we arrive at (4.5). Therefore, the S-matrix

$$R_{(12),3}(u) = P^+_{12} R_{12,3}(u) P^+_{12} \tag{4.8}$$

describes scattering of some spin -1 particle by the spin - 1/2 particle. The question arises if the S-matrix is factorized. Since from this moment several kinds of particles are involved, the factorization equation (3.2) must be understood now in a generalized sense as the equation in the space $V_1 \otimes V_2 \otimes V_3$ ($\dim V_1 \neq \dim V_2 \neq \dim V_3$). For example, in the case of scattering the spin - 1 particle (12) and two spin -1/2 particles 3 and 4 the factorization equation (3.2) reads

$$R_{(12),3}(u-v) R_{(12),4}(u) R_{34}(v) = R_{34}(v) R_{(12),4}(u) R_{(12),3}(u-v). \tag{4.9}$$

It turns out that $R_{(12),3}$ defined by (4.8) really satisfies (4.9). To see this, one needs to take the obvious equality

$$R_{12,3}(u-v) R_{12,4}(u) R_{34}(v) = R_{34}(v) R_{12,4}(u) R_{12,3}(u-v) \tag{4.10}$$

and multiply it by P^+_{12} from the right. With the use of the triangularity condition (4.5) and $(P^+_{12})^2 = P^+_{12}$ one obtains (4.9).

On considering in quite analogous way the scattering of two spin - 1 particles we obtain the S-matrix

$$R_{(12),(34)}(u) = P^+_{12} P^+_{34} R_{23}(u-\eta) R_{24}(u) R_{13}(u) R_{14}(u+\eta) P^+_{12} P^+_{34} . \tag{4.11}$$

As in the previous case, the S-matrix (4.11) can be shown to satisfy YBE. It is worth noticing that the YBB (4.11) coincides (up to a similarity transformation) with the one found previously in [19]. The YBB (4.11) is regular and, therefore, as discussed in Sec.3.2, can be considered as L-operator for some quantum chain model generalizing the XYZ-model to the spin – 1 case.

Let us stress that the S-matrix terminology is used here for the sake of clearness only. Of all properties of S-matrices we use only YBE and put away the properties of unitarity, crossing etc.

We proceed now to generalize the above construction to the case of arbitrary spin l. Our hypothesis is as follows. The YBB, corresponding to the scattering of the spin 1/2 particle by spin - - l particle is

$$R_{(1,2,\dots,2l),a}(u)=P^{+}_{1,2,\dots,2l}\,R_{1,2,\dots,2l;a}(u)\,P^{+}_{1,2,\dots,2l} \tag{4.12}$$

where

$$R_{1,2,\dots,2l;a}(u)=R_{1,a}\left(u-\tfrac{2l-1}{2}\eta\right)R_{2,a}\left(u-\tfrac{2l-3}{2}\eta\right)\dots R_{2l,a}\left(u+\tfrac{2l+1}{2}\eta\right)$$

and $P^{+}_{1,2,\dots,2l}$ is the projector (symmetrizer) onto the $(2l+1)$-dimensional subspace of the space $\overset{2l}{\underset{1}{\otimes}} C^2$. Quite analogous but more cumbersome formulas can be written for scattering of particles of arbitrary spin.

We can prove the factorization equation for the YBB (4.12) for l = 3/2 in the general XYZ-case and for arbitrary l in the degenerate XXX-case (2.75).

To prove YBE for (4.12) like in case l = 1, it is enough to check the triangularity condition

$$(I-P^{+}_{1,2,\dots,2l})\,R_{1,2,\dots,2l;a}(u)\,P^{+}_{1,2,\dots,2l}=0. \tag{4.13}$$

The condition (4.13) will apparently be satisfied if we succeed in representing the projector $1-P^{+}_{1,2,\dots,2l}$ as

$$I-P^{+}_{1,2,\dots,2l}=\sum_{j=1}^{2l-1}X_j\,P^{-}_{j,j+1} \tag{4.14}$$

where X_j are some operators. If (4.14) is true, then on substituting (4.12) and (4.14) into (4.13) and applying (4.7) many times we shall be able to bring all the projectors $P^{-}_{j,j+1}$ to the right side where they will annihilate with $P^{+}_{1,2,\dots,2l}$.

In the case l=3/2 the representation (4.14) reads

$$I-P^{+}_{1,2,3}=P^{-}_{12}+P^{-}_{23}+\frac{2}{3}P^{+}_{12}\mathcal{P}_{23}P^{-}_{12}+\frac{2}{3}P^{+}_{23}\mathcal{P}_{12}P^{-}_{23} \qquad (4.15)$$

where $\mathcal{P}_{12}$ and $\mathcal{P}_{23}$ are the permutation operators(3.4). In the case $l > 3/2$ the search for the representation (4.14) involves cumbersome calculations which we have not completed yet.

The problem becomes substantially simpler in the case of the XXX-model (2.75). The YBB (2.74) degenerates into the SU(2)-invariant operator bundle (2.75) ($2\varkappa=\eta$)

$$R_{12}(u)=u+\eta\mathcal{P}_{12} \qquad (4.16)$$

and there arises another value of the spectral parameter $u=\eta$ for which $R_{12}(u)$ turns into a projector:

$$R_{12}(\eta)=\eta(I+\mathcal{P}_{12})=2\eta P^{+}_{12}\,. \qquad (4.17)$$

The property (4.17) is readily generalized by induction to the higher order symmetrizers [39] :

$$P^{+}_{12\ldots n+1}=\frac{1}{\eta(n+1)}P^{+}_{12\ldots n}R_{1\,n+1}(n\eta)P^{+}_{23\ldots n+1}\,. \qquad (4.18)$$

Upon using the representation (4.18) and YBE one obtains

$$R_{1,2,\ldots,2l;a}(u)P^{+}_{1,2,\ldots,2l}=P^{+}_{1,2,\ldots,2l}R_{1,2,\ldots,2l;a}(u)P^{+}_{1,2,\ldots,2l} \qquad (4.19)$$

which proves (4.13).

The careful analysis of the above calculations shows that they are based essentially on two facts. The first is YBE. The second is that the YBB in question turns into a projector at a certain value of the spectral parameter. For generalizations of the above results to the case SU(N), see [39] . In the paper [39] the eigenvalues of the YBBs obtained are also calculated.

To finish the Subsection let us note that the triangularity of the proper combination of YBBs has been used in [71] for obtaining some functional equations for the partition function of certain models of lattice statistics.

A procedure of multiplying S-matrices which is close enough to the one described above is employed in [120] for calculating the bound state S-matrices in the framework of factorized S-matrix theory.

4.2. Higher spin ferromagnetic chains.

In the previous Subsection we have described a vast family of YBBs corresponding to arbitrary finite-dimensional representations of SU(2). One can prove [39] that the YBBs describing the scattering of equal spin particles are regular in the sense of (3.14). Therefore, by virtue of Subsection 3.2 they must generate some exactly soluble quantum chain models which can be considered as generalisations of XYZ model to higher spins.

Since it is not our aim to present here a detailed theory of the higher spin ferromagnets we shall restrict ourselves with several comments. For the sake of similicity we consider the isotropic (XXX) case.

So, consider the chain of N sites each carrying spin s. The space of quantum states is $\mathcal{H}_{\ell} = \overset{N}{\underset{1}{\otimes}} V_a$, $V_a = C^{2s+1}$. Let $R^{\ell s}$ be the YBB acting in the space $C^{2\ell+1} \otimes C^{2s+1}$. According to (3.15) we can identify $R^{\ell s}$ with an L-operator $L_n^{(\ell)}(u)$, the space $C^{2\ell+1}$ being considered as the auxiliary and C^{2s+1} as the quantum one. Due to YBB the traces of the corresponding transition matrices

$$t_{2\ell}(u) = \mathrm{tr}\, T_{2\ell}(u) = \mathrm{tr}\, L_N^{(\ell)}(u) L_{N-1}^{(\ell)}(u) \ldots L_1^{(\ell)}(u) \tag{4.20}$$

commute for arbitrary l's

$$[\, t_m(u), t_n(v)] = 0 . \tag{4.21}$$

Thus, we have a wide choice of L-operators of arbitrary matrix dimensions $(2\ell+1 = 2, 3, \ldots)$. According to (3.17-18) the local Hamiltonian can be extracted from $t_{2\ell}(u)$ if l=s. However, from the viewpoint of the algebrized Bethe ansatz the (2 x 2)-dimensional L-operator is more useful (l=1/2). To clear up the point, note that $L_n^{(1/2)}(u)$ (4.12) can be written [39] as

$$L_n^{(1/2)}(u) = u + \eta \sum_{\alpha=1}^{3} S_n^{\alpha} \sigma^{\alpha} = \begin{pmatrix} u + \eta S_n^3 & \eta S_n^- \\ \eta S^+ & u - \eta S_n^3 \end{pmatrix} \tag{4.22}$$

where $S_n^{\pm} = S_n^1 \pm i S_n^2$. The pseudovacuum $|0\rangle$ is defined then by $S_n^- |0\rangle = 0$, $\forall n$. On writing $T_1(u)$ as

$\begin{pmatrix} A(u) & B(u) \\ B^+(u) & A^+(u) \end{pmatrix}$ we can proceed along the same line as in Sec.2.2, the operator $B^+(u)$ ($B(u)$) being interpreted as creation (annihilation) operator of elementary excitations (magnons). Omitting the standard calculations (cf.Sec.2.2) we present here the final periodicity equation resulting like (2. 48) from the condition that the state $|u_1, u_2, \ldots, u_n\rangle = B^+(u_1)\ldots B^+(u_n)|0\rangle$ be an eigenstate of $t_1(u)$. The equation reads

$$\left(\frac{u_j - s\eta}{u_j + s\eta}\right)^N = \prod_{\substack{i=1 \\ i\neq j}}^{n} \frac{u_j - u_i - \eta}{u_j - u_i + \eta}, \quad j = 1, 2, \ldots, n. \qquad (4.23)$$

The corresponding eigenvalue $\nu_1(u)$ of $t_1(u)$ is

$$\nu_1(u) = (u - s\eta)^N \prod_{i=1}^{n} \frac{u - u_i + \eta}{u - u_i} + (u + s\eta)^N \prod_{i=1}^{n} \frac{u - u_i - \eta}{u - u_i}. \qquad (4.24)$$

However, what we really need is the eigenvalue $\nu_{2s}(u)$ of $t_{2s}(u)$ from which, as mentioned above, the eigenvalue of the local Hamiltonian can be obtained. Thus, the problem arises how to express $t_{2s}(u)$ (or, more generally, $t_{2\ell}(u)$) in terms of $t_1(u)$. The rest of the Section is devoted to a discussion of this problem.

The starting point of our reasoning is the equality

$$T_{2\ell}(u) = P^+_{1,2,\ldots,2\ell}\, T_1\left(u - \tfrac{2\ell-1}{2}\eta\right) \otimes \ldots \otimes T_1\left(u + \tfrac{2\ell-1}{2}\eta\right) P^+_{1,2,\ldots,2\ell} \qquad (4.25)$$

resulting from (4.11-12) and expressing $T_{2\ell}(u)$ in terms of $T_1(u)$. $P^+_{1,2,\ldots,2\ell}$ in (4.25) denotes the symmetrizer in the auxiliary space.

It is instructive to consider at first the classical case. Since the coupling constant η is assumed to be proportional to the Planck constant $\hbar$, (4.25) in the classical limit reads

$$T_{2\ell}(u) = P^+_{1,2,\ldots,2\ell}\, T_1(u) \otimes \ldots \otimes T_1(u)\, P^+_{1,2,\ldots,2\ell} \qquad (4.26)$$

Since (2 x 2)-dimensional matrix $T_1(u)$ has only two spectral invariants $t(u) = t_1(u) = \mathrm{tr}\, T_1(u)$ and $d(u) = \det T(u)$, all the traces $t_{2\ell}(u)$ must be expressible in terms of $t(u)$ and $d(u)$. A simple combinatorial calculation yields the answer

$$t_n(u)=\sum_{k=0}^{[n/2]}(-1)^k\binom{n-k}{k}d^k(u)\,t^{n-2k}(u), \tag{4.27}$$

For example,

$$\begin{aligned} t_1(u) &= t(u), \\ t_2(u) &= t^2(u)-d(u), \\ t_3(u) &= t^3(u)-2d(u)t(u). \end{aligned} \tag{4.28}$$

Returning to the quantum case we put forward the following hypothesis

$$\begin{aligned} t_n(u+\tfrac{n-1}{2}\eta) = {} & t(u)t(u+\eta)\ldots t(u+(n-1)\eta)- \\ & -d(u)t(u+2\eta)\ldots t(u+(n-1)\eta)- \\ & -t(u)d(u+\eta)t(u+3\eta)\ldots t(u+(n-1)\eta)-\ldots \\ & \ldots -t(u)t(u+\eta)\ldots t(u+(n-3)\eta)d(u+(n-2)\eta)+ \\ & +d(u)d(u+2\eta)t(u+4\eta)\ldots t(u+(n-1)\eta)+\ldots \end{aligned} \tag{4.29}$$

The summation in (4.29) is taken over all possible "pairings" in the product $t(u)t(u+\eta)\ldots t(u+(n-1)\eta)$. The "pairing" is understood here as replacing of two adjacent factors $t(u+k\eta)\times$ $\times\, t(u+(k+1)\eta)$ by the quantum determinant $d(u+k\eta)$. The quantum determinant $d(u)$ introduced first in [30] is defined (in case of 2 x 2 transition matrix) by

$$d(u)=\mathrm{tr}\, P^-_{12}\, T_1(u)\otimes T_1(u+\eta) \tag{4.30}$$

where P^-_{12} is the antisymmetrizer in the auxiliary space. More comments on quantum determinants can be found in Sec.5.

For small n's (4.29) reads

$$t_1(u)=t(u), \tag{4.31}$$

$$t_2(u)=t(u-\tfrac{\eta}{2})t(u+\tfrac{\eta}{2})-d(u-\tfrac{\eta}{2}), \tag{4.32}$$

$$t_3(u)=t(u-\eta)t(u)t(u+\eta)-d(u-\eta)t(u+\eta)-t(u-\eta)d(u). \tag{4.33}$$

It is an easy task to show that (4.29, 31-33) turn in the classical limit into (4.27-28).

We can prove the hypothesis (4.29) in cases n=1, 2, 3. For n=1 it is obvious. In case n=2 the proof is given by

$$\begin{aligned} t_2(u) &= \mathrm{tr}\, P^+_{12}\, T_1(u-\tfrac{\eta}{2})\otimes T_1(u+\tfrac{\eta}{2})\, P^+_{12} = \\ &= \mathrm{tr}(I-P^-_{12})\, T_1(u-\tfrac{\eta}{2})\otimes T_1(u+\tfrac{\eta}{2}) = \\ &= \mathrm{tr}\big(T_1(u-\tfrac{\eta}{2})\otimes T_1(u+\tfrac{\eta}{2}) - P^-_{12}\, T_1(u-\tfrac{\eta}{2})\otimes T_1(u+\tfrac{\eta}{2})\big) = \\ &= t(u-\tfrac{\eta}{2})\, t(u+\tfrac{\eta}{2}) - d(u-\tfrac{\eta}{2}) . \end{aligned} \tag{4.34}$$

In course of (4.34) we have used (4.5), (4.7), (4.30) and the cyclic property of the trace.

In quite a similar way the case n=3 is considered. On substituting the representation (4.15) of P^+_{123} in (4.26) and applying (4.5), (4.7), (4.30) we arrive at (4.33).

The general proof of the hypothesis which we believe to be true must be based on a generalisation of the representation (4.15) for the higher antisymmetrizers.

5. QUANTUM DETERMINANTS

In the present Section we shall study in detail the notion of the quantum determinant introduced in the previous Section. To state the problem let us consider first the classical case.

Let $L(x, u)$ be an ultralocal (cf.Sec.2.1) classical L-operator which satisfies the fundamental relation (2.11) with some r-matrix $r(u)$. We consider the general case of $L(x,u)$ being an $(n \times n)$-matrix. It follows immediately from (2.11) that not only the traces as in (2.40) but also traces of arbitrary power of the monodromy matrix $T(x_1, x_2; u)$ are in involution:

$$\{t^{(k)}(u), t^{(m)}(v)\}=0, \quad t^{(k)}(u)=tr\, T^k(x_1,x_2;u). \tag{5.1}$$

The quantities $t^{(k)}(u)$ are power sums of the eigenvalues of $T(u)$. We shall use also the symmetric functions $\sigma^{(m)}(u)$ of the same eigenvalues which correspond to the sums of the principal minors of the matrix $T(u)$ We define $\sigma^{(m)}(u)$ by

$$\sigma^{(m)}(u) = tr_{(m)} P^-_{1,2,\dots,m} \prod_{i=1}^{m} T_i(u) . \tag{5.2}$$

The subscript i in (5.2) shows in which of spaces V_i in the space

$$\mathcal{V}^{(m)} = \overset{m}{\underset{i=1}{\otimes}} V_i = V_1 \otimes V_2 \otimes \dots \otimes V_m , \quad V_i \simeq C^n \tag{5.3}$$

the matrix $T_i(u)$ acts nontrivially, i.e.

$$T_i(u) = \underbrace{I \otimes \dots \otimes I}_{i-1} \otimes T(u) \otimes I \otimes \dots \otimes I . \tag{5.4}$$

The operator $P^-_{1,\dots,m}$ in (5.2) is the projector (antisymmetrizer) onto the antisymmetric subspace of $\mathcal{V}^{(m)}$. The trace $tr_{(m)}$ in (5.2) is taken over the whole space $\mathcal{V}^{(m)}$. Apparently, $\sigma^{(1)}(u)=t^{(1)}(u)=tr\, T(u)$; $\sigma^{(n)}(u)=\det T(u)$.

Since $\sigma^{(m)}(u)$ can be expressed in terms of $t^{(k)}(u)$ according to Newton's formulas they also are in involution.

Due to (5.1) the quantities $t^{(k)}(u)$ (or $\sigma^{(m)}(u)$) can be used as generating functions of some integrals of motion for the model in question. In the general case the set $\{\sigma^{(k)}(u)$, $k=1, \dots$

... , n-1 } is the complete set of independent conserved quantities (cf., for example, the matrix NS [77, 33, 34]).

Proceeding to the quantum case it is natural to look for some quantum analogue of the quantities $\sigma^{(k)}(u)$. The extra conserved quantities are needed e.g. for applying QSTM to the matrix NS model whose classical Hamiltonian is contained in the family $\sigma^{(k)}(u)$ (k > 1, k=min(m , n), for m , n see (2.71)). It turns out that the form of the quantum generalization of $\sigma^{(k)}(u)$ depends essentially on the properties of the R-matrix which intertwines the quantum monodromy matrices (2.36). For the sake of simplicity, we restrict our exposition to the case of the simplest nontrivial R-matrix:

$$R(u) = u + \eta \mathcal{P} \tag{5.5}$$

where $\mathcal{P}$ is the permutation operator in $V \otimes V$ ($V \simeq C^n$). The R-matrix (5.5) serves a number of completely integrable models: NS (scalar n=2, vector, matrix [33, 34]), SU(N) isotropic Heisenberg ferromagnet [35] , nonabelian Toda chain [32] , the quantum N-wave problem and some others.

So, let the quantum monodromy matrix T(u) satisfy the fundamental relation (2.36) with the R-matrix (5.5). We define now the quantum symmetric function $\sigma^{(m)}(u)$ by

$$\sigma^{(m)}(u) = tr_{(m)} P^{-}_{1,2,\dots,m} T_1(u) T_2(u+\eta) \dots T_m(u+(m-1)\eta). \tag{5.6}$$

The notation in (5.6) is the same as in the classical case (5.2). It is easy to see that in the classical limit, $\eta \to 0$, (5.6) turns into (5.2). The main property of the quantum operators $\sigma^{(m)}(u)$ introduced by (5.6) is their commutativity

$$[\sigma^{(k)}(u), \sigma^{(m)}(v)] = 0 . \tag{5.7}$$

Due to space limitations we do not present here the complete proof of (5.7). The main idea is to prove that the monodromy matrices $\mathcal{T}^{(k)}(u)$ and $\mathcal{T}^{(m)}(v)$ defined by

$$\mathcal{T}^{(k)}(u) = P^{-}_{1,2,\dots,k} T_1(u) T_2(u+\eta) \dots T_k(u+(k-1)\eta) \tag{5.8}$$

satisfy the equation

$$\mathcal{R}^{(k,m)}(u-v) \mathcal{T}^{(k)}(u) \mathcal{T}^{(m)}(v) = \mathcal{T}^{(m)}(v) \mathcal{T}^{(k)}(u) \mathcal{R}^{(k,m)}(u-v) \tag{5.9}$$

with some R-matrix $\mathcal{R}^{(k,m)}(u)$. The proof of (5.9) is very much the same as the proof of YBE for the higher YBBs constructed in the previous Section (see also [39]). It uses the fundamental relation (2.36) and the representation of the antisymmetrizer $P^-_{1,2,\ldots,m+1}$ in terms of the R-matrix (5.5)

$$P^-_{1,2,\ldots,m+1} = \frac{-1}{\eta(m+1)} P^-_{2,3,\ldots,m+1} R_{1\,m+1}(-m\eta) P^-_{1,2,\ldots,m} \tag{5.10}$$

(cf. (4.18)).

In the rest of this Section we shall concentrate on the functional $\sigma^{(n)}(u)$. By analogy with the classical case it is natural to call the operator $\sigma^{(n)}(u)$ the quantum determinant of $T(u)$:

$$d(u) = \mathrm{Det}\,T(u) = \sigma^{(n)}(u) = \mathrm{tr}_{(n)} P^-_{1,2,\ldots,n} T_1(u)\ldots T_n(u+(n-1)\eta). \tag{5.11}$$

The properties of $d(u)$ are quite analogous to those of the classical c-number determinant:

1. $d(u)$ is invariant under the similarity transformation

$$T(u) \to U T(u) U^{-1}, \quad U \in GL(n,C).$$

2. The multiplicativity property. If $T(u)=T(u,1)T(u,2)$, and all the matrix elements of $T(u,1)$ commute with those of $T(u,2)$, then

$$d(u) = d_1(u)\, d_2(u), \quad d_i(u) = \mathrm{Det}\,T(u,i), \; i=1,2.$$

3. $$[\,d(u), d(v)\,] = 0 .$$

Moreover, in case of the R-matrix (5.5) $d(u)$ commutes with every matrix element of $T(u)$ and is scalar operator in $\mathcal{H}_q$.

4. $d(u)$ takes part in inverting $T(u)$

$$T^{-1}(u+(n-1)\eta) = \frac{n}{d(u)} \mathrm{tr}_{(n-1)} P^-_{1,\ldots,n} T_1(u)\ldots T_{n-1}(u+(n-2)\eta). \tag{5.12}$$

The first property is obvious. The second one follows from the fact that rank $P^-_{1,2,\ldots,n} = 1$ and the representation

$$T_1(u)\ldots T_n(u+(n-1)\eta) = T_1(u,1)T_2(u+\eta,1)\ldots T_n(u+(n-1)\eta,1)\,T_1(u,2)\ldots T_n(u+(n-1)\eta,2).$$

The third property follows from (5.9) for $m=n$, $k=1$, from

$$\mathcal{T}^{(n)}(u) = d(u)\, P^{-}_{1,2,\ldots,n} \tag{5.13}$$

and the fact that the quantum determinant of the R-matrix (5.5) is a scalar operator in V. To prove (5.12) one needs to multiply (5.13) from the right side by $T_n^{-1}(u+(n-1)\eta)$ and to take the trace over the first (n-1) spaces of the product $V_1 \otimes \ldots \otimes V_{n-1} \otimes V_n$ using the relation

$$tr_{(n-1)}\, P^{-}_{1,2,\ldots,n} = \frac{1}{n} I_n$$

(I_n being the unit operator in V_n).

Below we give two examples of quantum determinants for a linear problem on a chain. Due to the multiplicativity property the quantum determinant of $T_N(\ )$ is equal to the product of $L_k(u)$ operator determinants $d_k(u)$.

1. The auxiliary space is C^2 and quantum space is C^{2s+1}, operator $L_k(u)$ is given by (4.22)

$$d_k(u) = u(u+\eta) - \eta^2 \sum_{\alpha=1}^{3} (S_k^{\alpha})^2 . \tag{5.14}$$

2. The dimentions of auxiliary and quantum spaces are equal, the operator $L_k(u)$ coincides with simplest nontrivial R-matrix (5.5)

$$d_k(u) = u(u+\eta)\ldots(u+(n-2)\eta)(u+n\eta) . \tag{5.15}$$

References

1. Teoriya solitonov, ed. S.P.Novikov, Moskva, Nauka, 1980.
2. Ablowitz M., Studies in Appl.Math., 58, 17 (1978).
3. Solitons, Topics in Current Physics, 17 , eds. R.K.Bullough and P.J.Caudrey, Springer Verlag, 1980.
4. Mathematical Physics in One Dimension, eds. E.H.Lieb and D.C. Mattis, Academic Press, New York and London, 1966.
5. Phase Transitions and Critical Phenomena, eds. S.Domb and M.S. Green, Academic Press, New York and London, 1972.
6. Manakov S.V., Zakharov V.E., Lett.Math.Phys., 5, 247 (1981).
7. Zamolodchikov A.B., Zh.Eksper.Teor.Fiz., 79, 641 (1980). Zamolodchikov A.B., Comm.Math.Phys., 79, 489 (1981).
8. Faddeev L.D., Sklyanin E.K., Doklady AN SSSR, 243, 1430 (1978).
9. Sklyanin E.K., Doklady AN SSSR, 244, 1337 (1979).
10. Kulish P.P., Sklyanin E.K., Phys.Lett., 70A, 461 (1979).
11. Thacker H.B., Wilkinson D., Phys.Rev., D19, 3660 (1979).
12. Honerkamp J., Weber P., Wiesler A., Nucl.Phys., B152, 266 (1979).
13. Faddeev L.D., Sklyanin E.K., Takhtajan L.A., Teor.mat.fiz., 40, 194 (1979).
14. Faddeev L.D., Takhtajan L.A., Uspekhi mat.nauk, 34, 13 (1979).
15. Faddeev L.D., Soviet Sci.Reviews, Contemporary Math.Phys., C1, 107 (1980).
16. Faddeev L.D., Physica scripta, 23 (1981).
17. Thacker H.B., Rev.Mod.Phys., 53, 253 (1981).
18. Sklyanin E.K., Zapiski nauch.semin.LOMI, 95, 55 (1980).
19. Kulish P.P., Sklyanin E.K., Zapiski nauch.semin.LOMI, 95,129(1980).
20. Izergin A.G., Korepin V.E., Fizika elem.chastits, to be published.
21. Bethe H., Z.Phys., 71, 205 (1931).
22. Onsager L., Phys.Rev., 65, 117 (1944).
23. Baxter R.J., Ann.Phys., 70, 193 (1972). Baxter R.J., Ann.Phys., 70, 323 (1972).
24. Baxter R.J., Ann.Phys., 76, 1 (1973).
25. Korepin V.E., Teor.Mat.Fiz., 41, 169 (1979).
26. Korepin V.E., Comm.Mat.Phys., 76, 165 (1980).
27. Izergin A.G., Korepin V.E., Comm.Math.Phys., 79, 303 (1981).
28. Izergin A.G., Korepin V.E., Doklady AN SSSR, 259, 76 (1981).
29. Izergin A.G., Korepin V.E., Smirnov F.A., Teor.Mat .Fiz., 48(1981).
30. Izergin A.G., Korepin V.E., Lett.Math.Phys., 5, 199 (1981).
31. Izergin A.G., Korepin V.E., Vestnik Len.Gos.Univ., to be published.
32. Korepin V.E., Zapiski nauch.semin.LOMI, 101, 90 (1980).
33. Kulish P.P., preprint LOMI, R-3-79 (1979).
34. Kulish P.P., Doklady AN SSSR, 255, 323 (1980).
35. Kulish P.P., Reshetikhin, N.Yu., Zh.Eksper.Teor.Fiz., 80,214(1981).
36. Kulish P.P., Reshetikhin, N.Yu., Zapiski nauch.semin.LOMI, 101, 101 (1981).
37. Gerdjikov V.S., Ivanov M.I., Kulish P.P., preprint E2-80-882, JINR, Dubna (1980).
38. Kulish P.P., Lett.Math.Phys., 5, 191 (1981).
39. Kulish P.P., Reshetikhin N.Yu., Sklyanin E.K., Lett.Math.Phys., to be published.
40. Kulish P.P., Sklyanin E.K., Phys.Lett.A, to be published.
41. Takhtajan L.A., Zapiski nauch.semin.LOMI, 101, 158 (1981).
42. Takhtajan L.A., Faddeev L.D., Zapiski nauch.semin.LOMI, 109, 134 (1981).
43. Faddeev L.D., Takhtajan L.A., Phys.Lett.A., to be published.
44. Smirnov F., Doklady AN SSSR, to be published.
45. Tsyplyaev S.A., Teor.Mat.Fiz., 48, 24 (1981).

46. Bashilov Yu.A., Pokrovski S.V., Comm.Math.Phys ., 76, 129 (1980).
47. Bergknoff H., Thacker H.B., Phys.Rev.Lett., 42, 135 (1979).
Bergknoff H., Thacker H.B., Phys.Rev., D19, 3666 (1979).
48. Creamer D.B., Thacker H.B., Wilkinson D., Phys.Rev., D19, 3660 (1979).
49. Creamer D.B., Thacker H.B., Wilkinson D., Phys.Rev., D21, 1523 (1980).
50. Creamer D.B., Thacker H.B., Wilkinson D., Phys.Lett., 92B, 144 (1980).
51. Creamer D.B., Thacker H.B., Wilkinson D., FERMILAB-Pub-80/25-THY (1980).
Creamer D.B., Thacker H.B., Wilkinson D., FERMILAB-Pub-81/23-THY (1981).
52. Grosse H., Phys.Lett., 86B, 267 (1979).
53. Fowler M., Phys.Lett., 94B, 189 (1980).
54. Honerkamp J., preprint Univ.Freiburg THEP 80/6 (1980).
55. Babelon O., de Vega H.J., Vialle C.M., preprint PAR LPTHE 81/5 (1981).
56. Berezin F.A., Pokhil G.P., Finkelberg V.M., Vestnik Mos.Gos.Univ., 1, 21 (1964).
57. McGuire I.B., J.Math.Phys., 6, 432 (1965).
58. Lieb E.H., Liniger W., Phys.Rev., 130, 1605 (1963).
59. Faddeev L.D., Zakharov V.E., Funkz.anal. i pril., 5, 18 (1971).
60. Sklyanin E.K., preprint,LOMI-E-3-79 (1979).
61. Manakov S.V., private communication.
62. Kuznetzov E.A., Mikhailov A.V., Teor.Mat.Fiz., 30, 303 (1977).
63. Izergin A.G., Kulish P.P., Lett.Math.Phys., 2, 297 (1978).
Izergin A.G., Kulish P.P., Teor.Mat.Fiz., 44, 189 (1980).
64. Berezin F.A., Sushko V.N., Zh.Eksper.Teor.Fiz., 45, 1293 (1965).
65. Coleman S., Phys.Rev., D11, 2088 (1975).
66. McCoy B.M., Wu T.T., Scientia Sinica, 22, 1021 (1979).
67. Yang C.N., Yang C.P., J.Math.Phys., 10, 1115 (1969).
68. Takahashi M., Progr.Theor.Phys., 46, 401 (1971).
69. Takahashi M., Suzuki M., Prog.Theor.Phys., 48, 2187 (1972).
70. Fowler M., preprint univ.Virginia (1981).
Fowler M., Zotos X., preprint univ.Virginia (1981).
71. Stroganov Yu.G., Phys.Lett., 74A, 116 (1979).
72. Baxter R.J., Exactly solved models, Proceedings of the 1980 NUFFIC Summer School Enschede, The Netherlands.
73. Zamolodchikov A.B., Comm.Math.Phys., 69, 165 (1979).
74. Jimbo M., Miwa J., Mori Y., Sato M., Physica, 1D, 80 (1980).
75. Ruijsenaars S.N.M., Ann.Phys., 132, 328 (1981).
76. Lieb E.H., Mattis D.C., J.Math.Phys., 6, 304 (1965).
77. Manakov S.V., Zh.Eksper.Teor.Fiz., 65, 505 (1973).
78. Yang C.N., Phys.Rev.Lett., 19, 1312 (1967).
79. Sutherland B., Phys.Rev.Lett., 20, 98 (1968).
80. Johnson J.D., Krinsky S., McCoy B.M., Phys.Rev., A8, 2526 (1973).
81. Luther A., Phys.Rev., B14, 2153 (1976).
82. Lüscher M., Nucl.Phys., B117, 475 (1976).
83. Faddeev L.D., Korepin V.E., Phys.Rep., 42C (1978).
84. Zamolodchikov A.B., Zamolodchikov Al.B., Ann.Phys., 120, 253 (1979).
85. Kulish P.P., Tsyplyaev S.A., Teor.Mat.Fiz., 46, 172 (1981).
86. Shankar R., Witten E., Phys.Rev., D17, 2134 (1978).
87. Mikhailov A.V., Olshanetsky M.A., Perelomov A.M., Comm.Math.Phys., 79, 473 (1981).
88. Bukhvostov A.P., Lipatov L.N., Nucl.Phys., B180, 116 (1981).
89. Gross D., Neveu A., Phys.Rev., D10, 3235 (1974).
90. Shei S.-S., Phys.Rev., D14, 535 (1976).
91. Berg B., Weiz P., Nucl.Phys., B146, 205 (1978).
92. Belavin A.A., Phys.Lett., 87B, 117 (1979).
93. Arinshtein A.E., Phys.Lett., 95B, 280 (1980).

Arinshtein A.E., Yadernaya Fizika, 33, 551 (1981).
94. Andrei N., Lowenstein J.H., Phys.Rev.Lett., 43, 1698 (1979).
Andrei N., Lowenstein J.H., Phys.Lett., 90B, 106 (1980).
Andrei N., Lowenstein J.H., Phys.Lett., 91B, 401 (1980).
95. Dutyshev V.N., Zh.Exper.Teor.Fiz., 78, 1332 (1980).
96. Dashen R.F., Hasslacher B., Neveu A., Phys.Rev., D12, 2443(1975).
97. Neveu A., Papanicolaou N., Comm.Math.Phys., 58, 31 (1978).
98. Mikhailov A.V., Zakharov V.E., Comm.Math.Phys., 74, 21 (1980).
99. Karowski M., Thun H.J., preprint DESY 80/105 (1980).
100. Faddeev L.D., Semenov-Tian-Shansky M.A., Vestnik Len.Gos.Univ., 13, 81 (1977).
101. Cherednik I.V., Yadernaya Fizika, 33, 278 (1981).
102. Mikhailov A.V., Zakharov V.E., Zh.Eksper.Teor.Fiz., 74, 1953 (1978).
103. Polyakov A.M., Phys.Lett., 72B, 224 (1977).
104. Goldschmidt Y.Y., Witten E., Phys.Lett., 91B, 392 (1980).
105. Lieb E.H., Wu F.V., Phys.Rev.Lett., 20, 1445 (1968).
106. Ovchinnikov A.A., Zh.Eksper.Teor.Fiz., 57, 2137 (1969).
107. Filev V.M., Teor.Mat.Fiz., 33, 918 (1977).
108. Choy T.C., Phys.Lett., 80A, 49 (1980).
109. Reyman A.G., Semenov-Tian-Shansky M.A., Inventiones Math., 54, 81 (1979).
110. Gutzwiller M.C., Ann.Phys., 124, 347 (1980).
111. Bogoyavlenski O.I., Comm.Math.Phys., 51, 201 (1976).
112. Polyakov A.M., Nucl.Phys., B164, 171 (1979).
Bruschi M. et al., J.Math.Phys.,21, 2749 (1980).
113. Babbitt D., Thomas L., J.Math.Anal., 72, 305 (1979).
114. Reshetikhin N.Yu., private communication.
115. Belavin A.A., Funkz.anal. i pril., 14, 18 (1980).
116. Cherednik I.V., Doklady AN SSSR, 249, 1095 (1979).
Cherednik I.V., Teor.Mat.Fiz., 43, 117 (1980).
117. Krichever I.M., Funkz.anal. i pril., 15, 22 (1981).
118. Fateev V.A., Zamolodchikov A.B., Yadernaya Fizika, 32, 581(1980).
119. Fateev V.A., Yadernaya Fizika, 33, 1419 (1981).
120. Karowski M., Nucl.Phys., B153, 244 (1979).

THE INVERSE SCATTERING TRANSFORMATION

AND THE FUNCTIONAL INTEGRATION METHOD

H.J. de Vega

Laboratoire de Physique Théorique et
Hautes Energies, Université Pierre et
Marie Curie, Tour 16 - 1er étage,
4, place Jussieu, 75230 Paris Cedex 05,
France

Contents

I - INTRODUCTION

In Quantum Field Theory and in Classical Statistical Mechanics functional integrals like

$$Z = \int\!\!\int \mathcal{D}\phi(x) \, \exp - \frac{S[\phi(\cdot)]}{\hbar} \tag{1}$$

play a central role. In QFT $\hbar$ is Planck's constant and in statistical mechanics $\hbar$ stands for the absolute temperature.

The computation of these functional integrals with the aid of the inverse scattering transformation (IST) is the subject of these lectures.

We use the IST as a functional change of variables in (1) in order to simplify the integrand $e^{-S/\hbar}$. The new variables are the scattering data of an auxiliar linear problem. In this problem $\phi(x)$ plays the role of a potential.

We deal with euclidean functional integrals. Then, in one-dimensional cases, x stands for the imaginary time. We identify this imaginary time with the axis where scattering takes place in the auxiliar linear problem.

In many physically interesting cases one can find IST that simplify the action S (S becomes separable). In those cases one can say that the model is an "Integrable System".

S expressed in terms of suitable scattering variables (SV) can be used at two levels

a) To search for saddle points from

$$\frac{\delta S}{\delta(s.v.)} = 0 \tag{2}$$

This equation being simpler than $\frac{\delta S}{\delta \phi(x)} = 0$. Their solutions are instantons expressed in SV which can describe semi-classical behaviour of Z ($\hbar \to 0^+$ or 0^-)

b) To compute Z by integrating over the SV. Two problems immediately arise here :

i) what are the integrations bounds for the SV ?

ii) what is the functional measure in SV ?

Level b) is up to now mainly limitated to one dimensional problems. Integration over the SV has been performed in the separable jacobian approximation. That is, assuming that the measure of the SV is factorizable when expressed in terms of the SV themselves. We get in this way integral representations for the ground state energies for the N- dimensional anharmonic oscillator and for the quantum pendulum. These formulas contain a large amount of information . In particular they possess i) correct analytic properties in the coupling constant plane ii) exact imaginary parts near branch points ($\equiv$ large orders of perturbation theory), iii) tunnel effects [see III and IV].

In what concerns level a) the results are not limited to one dimensional theories (see I).The IST in the angular momentum allows to deal with rotationally invariant problems in higher spatial dimensions (see V and VI).

II - THE ACTION EXPRESSED IN SCATTERING VARIABLES

In order to apply inverse scattering methods to a given functional integral one needs a suitable IST [1] to express $e^{-S/\hbar}$ in terms of the corresponding SV. This is possible if one relates S with the Fredholm determinant of a linear operator $\hat{L} - \lambda$ or with the expansion of the log of this determinant in powers of λ^{-1}.

As explicit examples we shall discuss here the N -dimensional anharmonic oscillator [2, 3] and the euclidean non-linear sigma model [4].

These methods also apply to other problems like the pendulum[5] and Toda-like systems as well as to field theories like two-dimensional Gross-Neveu and Chiral Gross-Neveu models, $(\vec{\phi}^2)^2$ and the non-linear sigma model in three and four dimensions [6]

The action of the N-dimensional anharmonic oscillator reads

$$S[\vec{\phi}(.)] = \int_{-\infty}^{+\infty} dx \left[\frac{1}{2}\left(\frac{d\vec{\phi}}{dx}\right)^2 + \frac{\mu^2}{2}\vec{\phi}^{\,2} + \frac{g}{N}(\vec{\phi}^2)^2 \right] \tag{3}$$

where $\vec{\phi} = (\phi_1, \ldots, \phi_N)$ and g is the coupling constant. x is the imaginary time. It is possible to express this action directly in terms of SV at least for $N=2$ [5]. We find easier to recast first Z as a functional integral over a single function $\alpha(x)$. We use for this purpose the identity (Stratonovich transformation [7])

$$\exp\left[-\frac{g}{N}\int dx\,(\vec{\phi}^2)^2\right] = \int \mathcal{D}\alpha\, \exp\left[-\int dx\,\alpha(x)^2 - 2i\sqrt{\frac{g}{N}}\int dx\,\alpha(x)\,\vec{\phi}^2(x)\right]. \tag{4}$$

Upon placing Eqs (3),(4) into (1) one gets, after performing the integration over $\vec{\phi}$, which becomes Gaussian,

$$Z(N,g) = \frac{1}{Z_0}\int \mathcal{D}\alpha\, \exp\left\{-\frac{N}{2}\,\mathrm{Log}\det\left[-\frac{d^2}{dx^2}+\mu^2+4i\sqrt{\frac{g}{N}}\,\alpha(.)\right] - \int dx\,\alpha(x)^2\right\} \tag{5}$$

Here we have set $\hbar = 1$. The semi-classical limit corresponds to $g/N \longrightarrow 0$. Z_0 is such that $Z(N,0) = \exp -\left(\frac{N}{2}\cdot 2L\right)$
(We consider the system into a large box of length L)

This functional integral has a constant stationary point at

$$\alpha(x) = \alpha_0 = \frac{1}{4i}\sqrt{\frac{N}{g}}\,\zeta(g) \tag{6}$$

where $z(g)$ verifies

$$z^2 (\mu^2 + z) = 4g^2$$

Hence $z(g)$ is a three-valued function of g. In the physical sheet for $\mu^2 = 1 > 0$

$$z(g) = -\frac{1}{3} + \left[g + \sqrt{g^2 - \frac{1}{27}}\right]^{2/3} + \left[g - \sqrt{g^2 - \frac{1}{27}}\right]^{2/3} \quad (7)$$

and

$$\lim_{g \to 0} \frac{z(g)}{2g} = 1 .$$

For $\mu^2 = -1 < 0$ the physical sheet is defined by

$$\lim_{g \to 0} z(g) = 1 \quad (8)$$

and one obtains

$$z(g) = \frac{1}{3} - e^{\frac{2}{3}i\pi}\left[-g + \sqrt{-g^2 - \frac{1}{27}}\right]^{2/3} - e^{-\frac{2}{3}i\pi}\left[-g - \sqrt{-g^2 - \frac{1}{27}}\right]^{2/3} \quad (9)$$

The systematic expansion of the exponent in the integrand of (5) around this extremum gives the $1/N$ series[8]. For the ground state energy we get

$$E_G(N,g) = \frac{N}{2}\left[\sqrt{\mu^2 + z} - \frac{z^2}{8g}\right] - \lim_{L \to \infty} \frac{1}{2L} \log I(N,g) \quad (10)$$

where

$$I(N,g) = \int \mathcal{D}v \, \exp - \frac{N}{2} S_{eff}[v(.)] \quad (11)$$

$$S_{eff}[v(.)] = \mathrm{Log}\det\left(\frac{-\frac{d^2}{dx^2} + \mu^2 + z + v(.)}{-\frac{d^2}{dx^2} + \mu^2 + z}\right) - \frac{1}{8g}\int dx\, v(x)^2 - \frac{z}{4g}\int dx\, v(x) . \quad (12)$$

and where the following shift of the integration variable has been done :

$$\alpha(x) = \alpha_0 + \frac{1}{4i}\sqrt{\frac{N}{\mu g}}\; v(x) \quad , \qquad v(\pm\infty) = 0 \tag{13}$$

We wish to point out that an α - representation analogous to eq. (12) also holds for the non-$O(N)$-symmetric anharmonic oscillator with action [2]

$$S'[\vec{\phi}(\cdot)] = \int_{-\infty}^{+\infty} dx \left[\frac{1}{2}\left(\frac{d\vec{\phi}}{dx}\right)^2 + \frac{1}{2}\sum_{a=1}^{N} \mu_a^2\, \phi_a^2 + \frac{g}{N}\left(\vec{\phi}^{\,2}\right)^2 \right] .$$

Eq. (12) tells us that the linear Schrodinger operator

$$\hat{L} - \lambda = -\frac{d^2}{dx^2} + v(x) - k^2 \quad , \; \lambda = k^2 , \quad x \in \mathbb{R}$$

is here the natural choice in order to express S_{eff} in terms of SV.

Let us briefly recall the direct scattering problem for the one dimensional Schrödinger equation [1]

$$\left[-\frac{d^2}{dx^2} + v(x) \right] \psi(x,k) = k^2\, \psi(x,k) \tag{14}$$

where $v(\pm\infty) = 0$. More precisely we assume that

$$\int_{-\infty}^{+\infty} dx\, (1 + |x|)\, v(x) < \infty \tag{15}$$

Let $\psi(x,k)$ be a solution with unit ingoing amplitude from the right. We have asymptotically

$$\psi(x,k) \underset{x \to +\infty}{=} e^{-ikx} + r(k)\, e^{ikx}$$

$$\psi(x,k) \underset{x \to -\infty}{=} t(k)\, e^{-ikx}$$

$r(k)$ and $t(k)$ are the reflection and transmission coefficients respectively. They verify the unitarity condition $|r(k)|^2 + |t(k)|^2 = 1$.

$t(k)^{-1}$ is called the Jost function. It is an analytic function of k in $\mathrm{Im}\, k > 0$ and its zeros in that region are the eigenvalues of eq. (15). They are always purely imaginary and we denote them by $i\varkappa_\ell$ $(\ell = 1, \dots, N_B)$. The corresponding eigenfunctions, $\psi_\ell(x)$, are defined such that $\psi_\ell(x) \underset{x \to -\infty}{=} e^{\varkappa_\ell x}$

Their normalization coefficients are then

$$c_\ell = \left(\int_{-\infty}^{+\infty} dx \; \psi_\ell(x)^2 \right)^{-1}$$

The set

$$SV = \left\{ r(k),\ 0 \leqslant k < +\infty ;\ \varkappa_\ell , c_\ell ,\ \ell = 1,\dots, N_B \right\} \tag{16}$$

is called the "scattering data" because the potential $v(x)$ can be obtained from the SV through the Gelfand-Levitan-Marchenko equation This is a linear integral equation and it can be written as [1]

$$K(x,y) + \Omega(x+y) + \int_x^{\infty} K(x,y')\,\Omega(y'+y)\,dy' = 0 \tag{17}$$

for $x < y$

and

$$K(x,y) = 0 \qquad \text{for} \quad x > y$$

Here $K(x,y)$ is the unknown and Ω is given in terms of the SV through

$$\Omega(x) = \int_{-\infty}^{+\infty} \frac{dk}{2\pi} \, e^{ikx} \, r(k) + \sum_{\ell=1}^{N_B} c_\ell \, e^{-\varkappa_\ell x}$$

Once eq. (17) is solved from a given kernel Ω , v follows from

$$v(x) = -2 \frac{d}{dx} K(x,x)$$

There is a one to one correspondence between "potentials" $v(x)$ in eq. (14) and the SV defined by eq. (16) provided $v(x)$ satisfies eq. (15) and the SV verify $|r(k)| \leqslant 1$, $r(k) = O(1/k)$ for large k , $r(-k) = r(k)^*$; $r(0) = -1$ if $|r(0)| = 1$ and the $\varkappa_j , c_j$ are positive numbers [1d].

Moreover, the effective action (12) can be explicitly written in terms of SV

using the dispersion relation for the Jost function

$$F(k) = \prod_{j=1}^{N_B} \left(\frac{k - i x_j}{k + i x_j} \right) \exp\left(\frac{1}{2\pi i} \int_{-\infty}^{+\infty} \frac{dk'}{k - k'} \, \mathrm{Im}\left[1 - |r(k')|^2\right] \right) \qquad \mathrm{Im}\, k > 0 \tag{18}$$

It can be shown that the Jost function coincides with the Fredholm determinant of the respective differential operator $\hat{L} - \lambda$. In our case

$$F(k) = \det \left(\frac{-\frac{d^2}{dx^2} + k^2 + v(\cdot)}{-\frac{d^2}{dx^2} + k^2} \right) \tag{19}$$

The expansion of the logarithm of this identity in powers of k^{-1} gives rise to the trace identities. The first two read [1b]

$$\int_{-\infty}^{+\infty} dx \; v(x) = -4 \sum_{j=1}^{N_B} x_j - \frac{1}{\pi} \int_{-\infty}^{+\infty} dk \; \mathrm{Im}\left[1 - |r(k)|^2\right] \tag{20}$$

$$\int_{-\infty}^{+\infty} dx \; v(x)^2 = \frac{16}{3} \sum_{j=1}^{N_B} x_j^3 - \frac{4}{\pi} \int_{-\infty}^{+\infty} k^2 dk \; \mathrm{Im}\left[1 - |r(k)|^2\right] \tag{21}$$

Using eqs.(18)-(21) the effective action eq. (12) writes

$$S_{eff}[v] = -2 \sum_{j=1}^{N_B} F(\varkappa^2) - 2 \int_0^{\infty} k \frac{dk}{\pi} f(k) \log\left[1 - |r(k)|^2\right] \tag{22}$$

where

$$F(\varkappa^2) = \mathrm{ArgTh} \frac{\sqrt{1 + \gamma(g)}}{\varkappa} + \frac{i\pi}{2} + \frac{\varkappa^3}{3g} - \frac{\gamma \varkappa}{2g} \tag{23}$$

and

$$f(k) = -\frac{k}{4g} + \frac{\sqrt{1+\gamma}}{4k\left[1 + \gamma + k^2\right]} - \frac{\gamma}{8gk}$$

Eq. (2) for the effective action of the N-dimensional anharmonic oscillator exhibits two fundamental properties.

a) S_{eff} completely separates in terms of SV

b) S_{eff} is independent of the phase of the reflection coefficient $r(k)$ and of the normalization coefficients c_ℓ .

This is a hidden symmetry in eq.(12) explicitely put forward by the IST.

Saddle points follows by extremizing eq.(11) with respect to the SV

$$\frac{\delta S}{\delta r(k)} = 0 \qquad k \in R \; ; \quad \frac{\partial S}{\partial \varkappa_j} = 0 \quad , 1 \leq j \leq N_B$$

The first equation has a unique solution

$$r(k) = 0 \qquad \text{for all} \quad k$$

The second one gives

$$\varkappa^2 \left[\varkappa^2 - \frac{3\gamma + 2\mu^2}{2} \right] = 0 \tag{24}$$

In consequence, the non trivial solution expresses in SV as

$$\varkappa = \sqrt{\tfrac{3}{2}\gamma + \mu^2} \;, \quad N_B = 1 \;, \quad r(k) = 0 \quad k \in R \;, \; c \; \text{arbitrary} \tag{25}$$

It is easy to obtain this solution in the representation $v(x)$ from the Gelfand-Levitan-Marchenko equation (17), with the result

$$v_c(x) = - 2\varkappa^2 \operatorname{sech}^2(\varkappa x - \delta) \tag{26}$$

where $2\delta = \ln(c/2\varkappa)$. This is a solution of the non-linear and non-local equation

$$\langle x | \left[-\frac{d^2}{dx^2} + \mu^2 + \gamma + v(.) \right]^{-1} | x \rangle = \frac{v(x) + \gamma}{4g}$$

The present methods allows to solve a slightly more general equation [2]

$$\frac{1}{N} \sum_{a=1}^{N} \langle x | \left[-\frac{d^2}{dx^2} + \mu_a^2 + \gamma + v(.) \right]^{-1} | x \rangle = \frac{v(x) + \gamma}{4g}$$

The saddle point (26) is an instanton that rules the large orders [9] of the $1/N$ series for the anharmonic oscillator (for details see [2]) .

In order to study the nature of the $1/N$ expansion (for example for the ground state energy)

$$E_G(g,N) = \sum_{K=-1}^{+\infty} \frac{A_K(g)}{N^K} \qquad (27)$$

one is interested in the behaviour of $A_K(g)$ for large K. In this regime one has the Cauchy integral representation for the coefficients

$$A_K(g) = -\lim_{L\to\infty} \frac{1}{2L} \oint \frac{dN}{2\pi i} N^{K-1} \int \mathcal{D}v \; e^{-\frac{N}{2} S_{eff}[v]}$$

This integral is dominated for large K by the instanton (26) After computing the determinant of small fluctuations around $v_c(x)$ the integration over N can be performed with the result

$$A_K(g) \underset{K\to\infty}{=} \mu \left(\frac{2}{\pi}\right)^{3/2} \frac{\Gamma(K+1/2)}{\left[\rho(g)\right]^{K+\frac{1}{2}}} \frac{\sqrt{g}\,\left(\frac{3}{2}g+\mu^2\right)^{3/4}}{\sqrt{g}\left[\sqrt{\frac{3}{2}g+\mu^2}+\sqrt{g+\mu^2}\right]} + c.c. \qquad (28)$$

where

$$\rho(g) = -\frac{1}{3g}\sqrt{\mu^2+\frac{3}{2}g} + \frac{i\pi}{2} - \log\frac{\sqrt{g+\mu^2}+\sqrt{\frac{3}{2}g+\mu^2}}{\sqrt{g}} \qquad (29)$$

This formula can also be derived by an independent way in the $O(N)$ symmetric case [10].

III - FUNCTIONAL INTEGRATION THROUGH THE SCATTERING VARIABLES

Let us discuss now the use of SV as integration variables in Quantum Mechanics. As a concrete example we consider the N-dimensional anharmonic oscillator but this method also works for other systems [5].

One can consider directly the SV as new integration variables because they are in one-to-one correspondence with $v(x)$ under some regularity hypothesis. Then, the corresponding Jacobian should be computed. However, one can do better by noting that the change from $v(x)$ to SV given by Eq. (15) can be recast as a canonical transformation. This is shown in Ref. [11] where the canonically conjugated variables

$$v(x) \quad , \quad \pi(x) = \int_{-\infty}^{x} v(y)\, dy \tag{30}$$

are considered together with the associated Poisson bracket

$$\{\alpha, \beta\} = \int dx \left[\frac{\delta \alpha}{\delta v(x)} \frac{d}{dx}\left(\frac{\delta \beta}{\delta v(x)}\right) - (\alpha \leftrightarrow \beta) \right] \tag{31}$$

We can trivially introduce the variable $\pi(x)$ by rewriting our functional integral (1) as

$$\int \prod_x \left[dv(x)\, d\pi(x)\, \delta\left(\pi(x) - \int_{-\infty}^{x} v(y)\, dy\right)\right] e^{-\frac{N}{2} S_{eff}(v)} = \iint d\mu \; e^{-\frac{N}{2} S_{eff}(v)} \tag{32}$$

Then we change from the canonical pair (π, v) to the set defined by Faddeev and Zajarov [11]

$$p_j = \varkappa_j^2 \quad , \quad q_j = 2 \ln [i c_j \dot{F}(i\varkappa_j)] \tag{33}$$

$$P(k) = -\frac{k}{\pi} \ln[1 - |r(k)|^2] \quad , \quad Q(k) = \mathrm{Arg}\, [r(k) F(k)]$$

Here $F(k)$ stands for the Jost function of the Schrodinger equation (14) and

$$\dot{F}(i\varkappa) = \left.\frac{dF(k)}{dk}\right|_{k = i\varkappa}$$

These new variables are canonical for the Poisson bracket (31) i.e.,

$$\{Q(k), P(k')\} = \delta(k - k') \quad , \quad \{q_j, p_\ell\} = \delta_{j\ell} \tag{34}$$

all other Poisson brackets vanishing. Then, the Jacobian of the transformation

$$[\pi(x), v(x); x \in \mathbb{R}] \longrightarrow CSV \equiv [P(k), p_j, Q(k), q_j, k \in \mathbb{R}^+, \; 1 \leq j \leq N_B, N_B \in \mathbb{N}] \tag{35}$$

has unit value. Here CSV stands for "canonical scattering variables". The Jacobian is equal to one only for $v(x)$ sufficiently smooth which is not necessarily the case for the integration variable in a functional integral. Consequently one should expect that quantum fluctuations will modify the classical unit value of the Jacobian associated to (35).

The canonical transformation (35) mixes in a complicated way coordinates and momenta. At the quantum level momenta and coordinates become non commuting operators and then it is not clear how to define this transformation in an operational formalism. In the functional integral approach this kind of problems gives rise to a non unit jacobian. Only for contact canonical transformations the Jacobian can be computed by discretizing the functional integral[12]. In other words, one needs an ultraviolet regulator in order to cut off the high energy fluctuations of the field responsable of the Jacobian variations. A finite volume cut off can also be necessary. Of course a discretization procedure is not unique. One can write many different discrete systems all of which have the same continuous limit. The best thing to do would be to define the discretization such that $exp\left[-\frac{N}{2}S_{eff}\right]$ continues to be completely factorizable when expressed in terms of the discrete variables analogous to the CSV. One cannot know a priori if such discretization exists. However, completely integrable systems defined over infinite lattices are known.

We wish to note that we are using CSV in a different way than they have been used in the literature[1]. For us, they replace $v(x)$ in the functional integral, where x is the imaginary time. Usually, scattering variables depending implicitly on the (real) time are associated with fields that are functions of space and time.

Let us now discuss the integration bounds on the CSV corresponding to a $v(x)$ varying between $-\infty$ and $+\infty$ for all finite x and subject to the boundary condition

$$v(\pm\infty) = 0 \tag{36}$$

In this case the eigenvalues χ_j^2 can take any positive value and the reflection coefficient $r(k)$ any complex value within the unit circle. Then, from Eq. (33) it follows that

$$0 \leqslant p_j < +\infty \,, j=1,\dots,N_B \quad ; \quad 0 \leqslant P(k) < +\infty \quad , \quad k \in \mathbb{R}^+ \qquad (37)$$

The number of bound states N_B being arbitrary, one must sum over N_B from zero to infinity. The variable q_j can take, in principle, any real value from $-\infty$ to $+\infty$. In fact, this infinity can be shown to be proportional to the infinite length of the imaginary time axis as follows. Suppose one does a translation $x \to x + X$. This changes $v(x)$ into $v(x+X)$ and does not affect p_j and $P(k)$. On the other hand, $r(k)$ and c_j transform like

$$c_j \to c_j \, e^{-x_j X} \quad , \quad r(k) \longrightarrow r(k) \, e^{2ikX}$$

because of the asymptotic behavior of the bound state wave functions. Hence q_j and $Q(k)$ change as

$$q_j \to q_j - 2\sqrt{p_j}\, X \quad , \quad Q(k) \to Q(k) + 2kX$$

If we put the system into a large box of length $2L$ we have

$$2\sqrt{p_j}\, L > q_j > -2\sqrt{p_j}\, L \;, \quad 2kL > Q(k) > -2kL \qquad (38)$$

We now proceed to discretize the functional integral (1) as well as the SV without attempting to exactly preserve the canonical nature of the mapping $(\pi, v) \longrightarrow CSV$ on the lattice. This will introduce a non unit Jacobian.

We shall assume that this Jacobian factorizes when expressed in terms of CSV. We call that the "separable Jacobian approximation" (SJA). This approximation means that we suppose that there are regions of the functional domain of integration where the Jacobian separates are that they are dominant. This is true in the semiclassical regime. The SJA as well as its consequences will be explicitly verify a posteriori (see IV).

We discretize the imaginary time axis as a lattice of $2M$ points $(2M \gg 1)$ over a length $2L$. That is $\Delta = L/M$ is the lattice spacing. The scattering variables become

$$Q_\alpha = \sqrt{\frac{\pi}{L}}\, Q(k_\alpha) \qquad , \qquad P_\alpha = \sqrt{\frac{\pi}{L}}\, P(k_\alpha) \tag{39}$$

together with the (q_j, p_j). Here $k_\alpha = \pi\alpha/L$ and α is an integer between $-M$ and $+M$. Here the distinction between "continuum" and "bound state" variables disappears. One should integrate over $2M - N_B$ pairs (Q_α, P_α) for each N_B, because the total number of independent variables is $2M$.

In conclusion we write our integration measure as

$$d\mu = \frac{1}{N_B!} \prod_{\alpha=1}^{2M-N_B} J_\alpha\, dQ_\alpha\, dP_\alpha \prod_{j=1}^{N_B} J_j\, dq_j\, dp_j \tag{40}$$

Here we write the Jacobian in the SJA as

$$J = \prod_{\alpha=1}^{2M-N_B} J_\alpha \prod_{j=1}^{N_B} J_j \tag{41}$$

The integration over all $v(x)$ and $\pi(x)$ corresponds in the CSV to sum over N_B from zero to $2M$ $(\gg 1)$ and to integrate over (P, Q, p_j, q_j) in the intervals (37) and (38). We are integrating independently over N_B different eigenvalues $p_j = \varkappa_j^2$ and then we must divide by $N_B!$ in order to avoid double counting of configurations.

We have now enough tools to integrate over the scattering variables. We use in eq. (32) the integration measure given by eq. (40) and eqs. (22)-(23) for S_{eff}. reads

$$I(N,g) = \lim_{M\to\infty} \frac{1}{Z_0} \sum_{N_B=0}^{2M} \frac{1}{N_B!} \prod_{\alpha=1}^{2M-N_B} \int_0^\infty \int_{-2Lk_\alpha}^{+2Lk_\alpha} \frac{\pi}{2L}\, dP(k_\alpha)$$

$$dQ(k_\alpha)\, e^{-NP(k_\alpha)Q(k_\alpha)} J_\alpha \prod_{j=1}^{N_B} \int_0^\infty \int_{-2\sqrt{p_j}L}^{+2\sqrt{p_j}L} J_j\, dp_j\, dq_j \exp[NF(p_j)] \, . \tag{42}$$

All integrals are here elementary except the one over p_j. Then, we obtain

$$I(N,g) = \lim_{M\to\infty} \frac{1}{Z_0}\left(\frac{2\pi}{N}\right)^M \sum_{N_B=0}^{2M} \frac{1}{N_B!}\left(\frac{2NL}{\pi}\right)^{N_B}\left[\int_0^\infty dp\,\sqrt{p}\,e^{NF(p)} J(p)\right]^{N_B} \prod_{\alpha=1}^{2M-N_B} \frac{k_\alpha J_\alpha}{f(k_\alpha)}$$

Furthermore in the $\Delta \to 0$, $M \to \infty$ limit

$$\prod_{\alpha=1}^{2M-N_B} \frac{J_\alpha k_\alpha}{f(k_\alpha)} = \exp\left\{-\frac{L}{\pi}\int_{-\pi/\Delta}^{+\pi/\Delta} dk \log\left[\frac{f(k)}{k J_k}\right]\left[1 + O\left(\frac{1}{L}\right)\right]\right\} \quad (43)$$

and we get for the ground state energy

$$E_G(N,g) = \frac{N}{2}\left[\sqrt{\mu^2+\gamma} - \frac{\gamma^2}{8g}\right] + \lim_{\substack{\Delta\to 0\\ L\to\infty}}\left\{\int_{-\pi/\Delta}^{\pi/\Delta}\frac{dk}{2\pi}\,\mathrm{Log}\,\frac{f(k)}{k J_k}\right.$$

$$\left. - \frac{1}{2L}\log\left[Z_0\left(\frac{N}{2\pi}\right)^{2M}\right]\right\} + \frac{N}{\pi}\int_0^\infty \sqrt{p}\,dp\,J(p)\,e^{NF(p)}.$$

The limit $\Delta \to 0$ of the expression within braces in eq. (43) is seen to be zero from the relation

$$\int_{-\infty}^{+\infty} dk\ \mathrm{Log}\left[\frac{4g}{k} f(k)\right] = 0 \quad (44)$$

and the definition of Z_0. This shows that in the SJA

$$J_k = \frac{1}{4g} \quad (45)$$

We then get

$$E_G(N,g) = \frac{N}{2}\left[\sqrt{\mu^2+\gamma} - \frac{\gamma^2}{8g}\right] + \frac{2N}{\pi}\int_0^\infty ds\, e^{NF(s)}\, s^2 J(s) + C_N(g) \quad (46)$$

where the term $C_N(g)$ accounts for higher order corrections to the SJA and we have set $s = \sqrt{p}$ in the integral.

IV - THE GROUND STATE ENERGY OF THE N DIMENSIONAL ANHARMONIC OSCILLATOR IN THE SEPARABLE JACOBIAN APPROXIMATION

The evaluation of the ground state energy of the N- component anharmonic oscillator led to eq. (46).

In this section we shall use the known large order behavior of the $1/N$ expansion of this ground state energy to determine as far as possible the Jacobian in the SJA.

The $1/N$ expansion of $E_G(N,g)$ has the form

$$E_G(N,g) = \sum_{K=-1}^{+\infty} A_K(g)\, N^{-K} \tag{47}$$

The coefficients $A_K(g)$ follow from eq. (46)

$$A_K(g) = \oint \frac{dN}{2\pi i} N^{K-1} E_G(N,g) = +\frac{2}{\pi} \oint \frac{dN}{2\pi i} N^K \int_0^\infty s^2 ds\, J(s)\, e^{NF(s)} \tag{48}$$

where we assume that $C_N(g)$ does not contribute for large K. The results will confirm this assumption.

The integral over s has a stationary point where

$$F'(s) = 0$$

This gives

$$s = s_0 \equiv \sqrt{\mu^2 + \frac{3}{2}\gamma} \tag{49}$$

This is precisely the bound state eigenvalue that defines in SV the instanton for the large orders on $1/N$ in the original functional integral [see sec. II].

We obtain from (48) by the steepest descent method

$$A_K(g) = -\frac{2}{\pi} J(s_0) \frac{\sqrt{g\gamma s_0}}{\sqrt{2\pi}} \frac{\Gamma(K+\frac{1}{2})}{F(s_0)^{K+1/2}} \left[1 + O\left(\frac{1}{K}\right)\right] . \tag{50}$$

We note that the leading factor $\Gamma(K+\frac{1}{2})\,F(s_0)^{-K-\frac{1}{2}}$ in Eq. (28) is correctly reproduced by this expression. The identification of (50) with the actual behavior (28) of $A_K(g)$ gives us the constraint

$$J(s_0) = \frac{2s_0^3}{g}\,\frac{1}{[s_0+\sqrt{\mu^2+z}\,]^2} \tag{51}$$

In other words, by fixing the function $J(s)$ on the curve $s=s_0(g)$ we obtain that eq. (4 6) for $E_G(N,g)$ also re-produces the determinant of small flucturations around the instanton of Ref 2. Moreover Eq. (51) suggests

$$J(s) = \frac{2s^3}{g}\,\frac{1}{[s+\sqrt{\mu^2+z}\,]^2} \tag{52}$$

for <u>all</u> s and g. In the following section we shall see that this last conjecture is correct.

As we see, the assumption that $C_N(g)$ does not contribute to the large order behavior of the $1/N$ series is consistent.

Finally, the integral representation for the ground state energy reads

$$E_G(N,g) = \frac{N}{2}\Big[\sqrt{\mu^2+z} - \frac{z^2}{8g}\Big] + I_N(g) + C_N(g)$$

$$I_N(g) = \frac{4N}{\pi g\mu^2}\int_0^{\infty\cdot e^{i\alpha}} \frac{s^5\,ds}{[s+\sqrt{\mu^2+z}\,]^2}\left[\frac{\sqrt{\mu^2+z}+s}{\sqrt{\mu^2+z}-s}\right]^{\frac{N}{2}} \exp\frac{N}{g}\Big(\frac{s^3}{3}-\frac{zs}{2}\Big) \tag{53}$$

This last formula contains a large amount of information about E_G.

1) For each complex g there are three inequivalent integration paths in the s-plane for which $I_N(g)$ converges. The integral will converge if s approaches infinity as

$$s = |s|\,e^{i\alpha}\;,\quad |s|\to\infty\;,\quad \alpha \text{ fixed}$$

and

$$0 > \mathrm{Re}\,(s^3/g) = -\,|s^3/g|\cos(3\alpha-\varphi)$$

where $g = |g| e^{i\varphi}$. This condition holds if

$$\pi/6 + \varphi/3 < \alpha < \pi/2 + \varphi/3$$

$$5\pi/6 + \varphi/3 < \alpha < 7\pi/6 + \varphi/3$$

$$3\pi/2 + \varphi/3 < \alpha < 11\pi/6 + \varphi/3$$

Hence $I_N(g)$ has a three-sheeted Riemann structure with branch points at $g = 0$ and infinity.

2) For large g and fixed N , $I_N(g) \sim g^{1/3}$

3) The discontinuity of $E_G(N, g)$ across its cut on $Re\, g < 0$ follows from eq. (46). This discontinuity for $0 > g > -(27)^{-1/2}$ comes solely from the singular factor $[\sqrt{\mu^2 + g} - s]^{-N/2}$ in the integrand. Thus, for $\mu^2 = 1$

$$\underset{0 < h < (27)^{-1/2}}{\mathcal{I}m\, E_G(N, -h + i0)} = \frac{4N}{\pi h} \sin\left(\frac{N\pi}{2}\right) \int_{\sqrt{1+g}}^{\infty} \frac{s^5\, ds}{[s + \sqrt{1+g}\,]^2} \left[\frac{s + \sqrt{1+g}}{s - \sqrt{1+g}}\right]^{N/2}$$

$$\exp\left\{-\frac{N}{h}\left[\frac{s^3}{3} - \frac{s g}{2}\right]\right\} \tag{54}$$

where we assume that $\mathcal{C}_N(g)$ does not contribute. This integral is dominated for $h \to 0^+$ by its lower bound .

Then we set

$$s = \sqrt{1+g}\, (1 + h t) \quad . \tag{55}$$

After some work we find

$$\mathcal{I}m\, E_G(N, -h + i0) = \left(\frac{2N}{h}\right)^{N/2} \frac{e^{-\frac{N}{3h}}}{\Gamma(\frac{N}{2})} \left\{ 1 + \sum_K T_K(N) \left(\frac{h}{N}\right)^K \right\} , \tag{56}$$

where the $T_K(N)$ are polynomials in **N** of degree 2K

$$T_K(N) = \sum_{j=0}^{2K} m_{j,K}\, N^{2K-j} \tag{57}$$

We obtain

$$m_{0,K} = \frac{(-1)^K}{K!} \frac{7^K}{2^{3K}} \; ; \; m_{1,K} = \frac{(-1)^{K-1}}{(K-1)!} \frac{7^{K-2}}{2^{3K-1}} (41K-104) \quad (58)$$

As it is well known, for $g < 0$ the oscillator becomes unstable and then the energy acquires an imaginary part. For $\mu^2 > 0 > g$, $X = 0$ is an unstable minimum because of tunnel effect (see fig. 1) and indeed $y_m(N, -h)$ is dominated for $h \longrightarrow 0^+$ by the instanton that goes through the potential barrier.[13,9] The action of this classical solution coincides with the exponent in eq. (56) and the determinant of small fluctuations (one loop) around it with the remaining factor in front of the braces in eq. (56)[13, 9]. Moreover $m_{0,K}$ and $m_{1,K}$ that correspond to contributions from diagrams with $(K+1)$ loops, exactly coincide with the available numerical values[14].

It is remarkable that the simple integral (53) can reproduce all that.

4) $I_N(g)$ has a saddle point for large N at

$$s_0 = \sqrt{\frac{3}{2} g + \mu^2} \quad (59)$$

This saddle point gives by steepest descent the large orders in the $1/N$ expansion [eq. (50)] .

In the case $\mu^2 = -1$ the contribution of this saddle point is generically complex

$$\epsilon = \pm 2i \sqrt{\frac{2Ng}{\pi g}} \frac{\left[\frac{3}{2} g - 1\right]^{3/4}}{\left[\sqrt{3g/2 - 1} + \sqrt{g - 1}\right]^2} \left[\frac{\sqrt{g-1} - \sqrt{3g/2 - 1}}{\sqrt{g-1} + \sqrt{3g/2 - 1}}\right]^{N/2}$$

$$\exp\left[-\frac{N}{3g}\sqrt{\frac{3g}{2} - 1}\right] \left\{1 + O\left(\frac{g}{N}\right)\right\} \quad (60)$$

This contribution becomes real for $N = 1$ and small positive g. In that case one gets from eq (8) and (60) [15]

$$\epsilon = \pm \frac{2^{3/4}}{\sqrt{\pi g}} e^{-\frac{1}{3\sqrt{2} g}} \left[1 + O(g)\right] \quad (61)$$

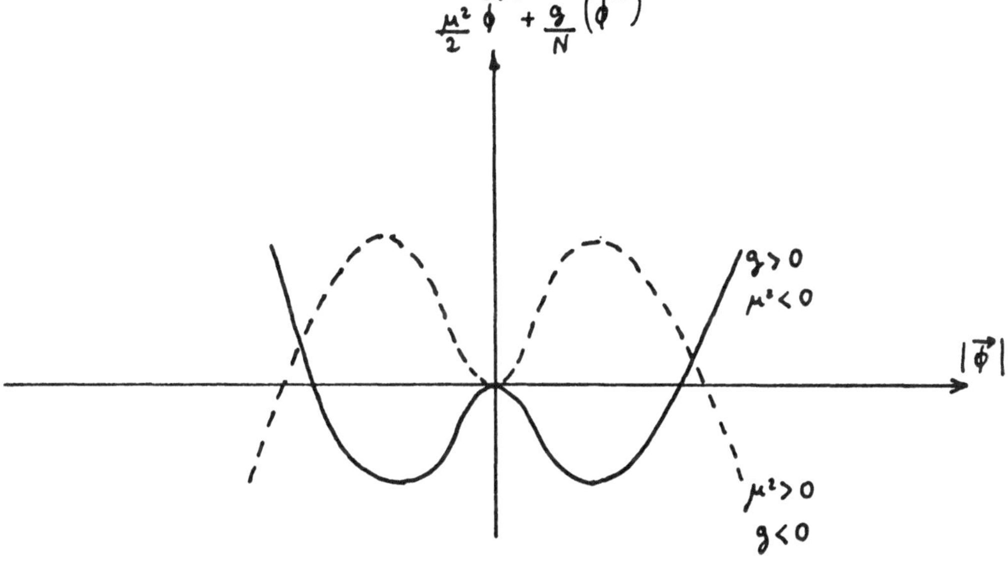

fig. 1

This exactly coincides with the tunneling contributions to the ground state and first excited state of the double well [16]. This means that eq. (46) contains the information about the instanton connecting the two degenerate minima for $\mu^2 < 0$ [Fig. 1] and about the small fluctuations around it.

In conclusion we see that eq. (53) contains most of what we know about the ground state of the anharmonic oscillator for both signs of μ^2.

We have discussed up to now the ground state energy. One can deal also with Green's functions.

The two point function of the N - component anharmonic oscillator reads
$$\int \mathcal{D}\vec{\phi}\ \phi_a(x)\,\phi_b(x')\, e^{-S[\vec{\phi}]} = \delta_{ab}\, G(x-x')$$

where $1 \leq a, b \leq N$ and $S[\vec{\phi}]$ is given by eq. (3).

In the α-representation we have
$$G(x-x') = \iint \mathcal{D}v\ \ g(x,x';v)\, e^{-\frac{N}{2} S_{eff}[v(.)]} \tag{62}$$
where Seff is given by eqs.(22)-(23) and
$$g(x,x';v(.)) = \langle x | \left[-\frac{d^2}{dx^2} + \mu^2 + \gamma + v(.) \right]^{-1} | x' \rangle$$

On the other hand, the usual expansion of Green's function in the eigenstates of the Hamiltonian gives
$$\tilde{G}(E) = 2 \sum_n \frac{E_n}{E^2 + E_n^2}\ \ |\langle 0 | \phi_a | n \rangle|^2 \tag{63}$$
where
$$\tilde{G}(E) = \int_{-\infty}^{+\infty} e^{iEx}\ G(x)\, dx \tag{64}$$
and $|n\rangle$, E_n stand for the eigenstates and eigenvalues of the anharmonic oscillator. Hence excited states of the system can be explicitly found if we succeed in computing the functional integral (62). In order to do that within the inverse scattering approach, we should express $g(x,x';v(.))$ in terms of the

CSV . $g(x,x';v(.))$ is related to the Jost solutions $f_\pm(k,x)$ of the auxiliary Schrödinger equation (14) as follows ;

$$g(x,x';v(.)) = \frac{f_+(im, x_>)\; f_-(im, x_<)}{2m\; F(im)} \qquad ; m = \sqrt{\mu^2 + z} \tag{65}$$

Here $F(k)$ stands for the Jost function which is immediately expressable in terms of CSV . The Jost solutions can be related to the CSV through a linear integral equation following Marchenko's formalism[1]. The kernel of this integral equation has a simple expression in CSV . In this way, the computation of the Green's function of the anharmonic oscillator reduces to a linear problem.

V - HIGHER DIMENSIONAL PROBLEMS : THE IST IN THE ANGULAR MOMENTUM

The IST in the angular momentum allows to deal with rotationally invariant problems in a number of spatial dimensions ν higher than one.

This IST is defined by linear equations where the spectral parameter λ is the angular momentum

$$\left[-\frac{d^2}{dr^2} - \frac{1}{4r^2} + v(r) - E \right] \Psi(\lambda, r) = -\frac{\lambda^2}{r^2}\, \Psi(\lambda, r) \tag{66}$$

(ν-dimensional radial Schrödinger equation)

$$\begin{pmatrix} \frac{d}{dr} & -v(r) + w(r) - 1 \\ v(r) + w(r) + 1 & -\frac{d}{dr} \end{pmatrix} \phi(\lambda, r) = \frac{\lambda}{r}\, \phi(\lambda, r) \tag{67}$$

where

$$\phi(\lambda, r) = \begin{pmatrix} \Psi_1 \\ \Psi_2 \end{pmatrix}$$

(two-dimensional radial Dirac equation)

We shall choose length units such $|E| = 1$.(We work at fixed "energy").

The scattering variables (SV) of these IST are natural variables to express renormalized determinants like

$$\operatorname{Log}\operatorname{Det}\left(\frac{-\partial^2 - E + v(.)}{-\partial^2 - E}\right) \quad - \text{counterterms} \tag{68}$$

or

$$\operatorname{Log}\operatorname{Det}\left(\frac{\not\partial + m - v(.) - \gamma_5 w(.)}{\not\partial + m}\right) \quad - \text{counterterms} \tag{69}$$

for rotationally invariant v , w which vanish at $r = \infty$. This kind of determinants appears in the effective action at the one loop level. An effective action of this kind generates the $1/N$ expansion in theories with an N-component field like the non-linear sigma model and the $(\vec{\Phi}^2)^2$ model. In usual perturbative expansions one finds such determinants in the generating functional of $1PI$ functions (also called effective action) at the one-loop order. The IST in the angular momentum helps to find saddle points of such non-local actions as well as to study the strong field behaviour.

Euclidean QFT corresponds to negative E in eqs.(68)-(69). Positive E is found for minkowskian QFT where v is time independent in eq.(68).

We shall deal here with the direct and inverse scattering problems for Schrödinger equation (66)[17]. The Dirac type equation (67) can be treated in a similar way.

Let us now define a regular solution of eq (66) by the boundary at $r = \infty$

$$\lim_{r \to \infty} e^{r} \varphi(\lambda, r) = 1 . \tag{70}$$

$\varphi(\lambda,r)$ is an entire function of λ^2 because the boundary condition is independent of λ . It is convenient to introduce Jost-type solutions by the boundary condition at $r = 0$

$$\lim_{r \to 0} r^{-\lambda-\frac{1}{2}} f(\lambda, r) = 1 \tag{71}$$

The scalar product associated to the Schrödinger eq.(66) is

$$(\chi_1,\chi_2) = \int_0^\infty \frac{dr}{r^2}\ \chi_1(r)^*\ \chi_2(r) \tag{72}$$

We shall assume that

$$\int_0^\infty r\ |v(r)|\ dr < \infty \tag{73}$$

The Jost function is given by the wronskian

$$F(\lambda) = \frac{1}{2}\ W[\varphi(\lambda,r),\ f(\lambda,r)] \tag{74}$$

Another expression for $F(\lambda)$ is

$$F(\lambda) = \lim_{r\to\infty} e^{-r} f(\lambda,r) \tag{75}$$

It follows that $F(\lambda)$ is analytic in the right half plane and

$$F(\lambda^*) = F(\lambda)^*$$

For the free equation $(v \equiv 0)$ the solutions can be expressed in terms of cylindric functions

$$\varphi_0(\lambda,r) = \sqrt{\frac{2r}{\pi}}\ K_\lambda(r) \quad ; \quad f_0(\lambda,r) = 2^\lambda\,\Gamma(\lambda+1)\sqrt{r}\ I_\lambda(r) \tag{76}$$

$$F_0(\lambda) = (2\pi)^{-1/2}\ 2^\lambda\ \Gamma(\lambda+1)$$

In general, there is a continuous spectrum plus a discrete one. The first is formed by the regular solutions for purely imaginary λ, $\lambda = i\tau$, $0 \leq \tau < +\infty$. For small distances

$$\varphi(i\tau,r) \underset{r\to 0}{=} \sqrt{\frac{2r}{\tau\,\mathrm{sh}\,\pi\tau}}\ D(\tau)\ \sin\left[\tau \ln\frac{r}{2} - \delta(\tau) - \mathrm{Arg}\,\Gamma(1+i\tau)\right] \tag{77}$$

Here

$$\frac{F(i\tau)}{F_0(i\tau)} = D(\tau)\ e^{i\delta(\tau)} \tag{78}$$

For each root λ_K of $F(\lambda)$ in $\mathrm{Re}\,\lambda > 0$ there is a normalizable eigenfunction. They form the discrete spectrum. It can be shown that all the λ_K are real and simple

$$F(\lambda_K) = 0 \qquad K = 1,2,\ldots, M$$

$$\varphi_K(r) \equiv \varphi(\lambda_K, r)$$

the normalization coeffieients are defined as

$$C_K = \left(\int_0^\infty \frac{dr}{r^2} \varphi_K(r)^2 \right)^{-1} \tag{79}$$

The set formed by $\{ \varphi(i\tau, r), \tau \in \mathbb{R}^+ ; \varphi_K(r), K = 1,\ldots, M \}$ is orthogonal and complete.

Let us now consider the inverse problem. In the present case the Gelfand Levitan equation turns to be [17]

$$K(r,r') + \Omega(r,r') + \int_r^\infty \frac{ds}{s^2} K(r,s)\, \Omega(s,r') = 0 \text{ for } r<r' \tag{80}$$

$$K(r,r') = 0 \qquad \text{if} \quad r > r'$$

Here the triangular kernel $K(r,r')$ transforms wave functions for a potential $v(r)$ into wave functions for a potential $\tilde{v}(r)$. K is the unknown in eq. (80) whereas $\Omega(r,r')$ is given by

$$\Omega(r,r') = \int_0^\infty \frac{\tau^2 d\tau}{2\pi} \varphi(i\tau, r)\, \varphi(i\tau, r') \left[|\tilde{F}(i\tau)|^{-2} - |F(i\tau)|^{-2} \right]$$

$$+ \sum_{K=1}^{\tilde{M}} \tilde{C}_K\, \varphi(\tilde{\lambda}_K, r)\, \varphi(\tilde{\lambda}_K, r') - \sum_{K=1}^{M} C_K\, \varphi(\lambda_K, r)\, \varphi(\lambda_K, r') \tag{81}$$

Here the quantities with (without) twiggle refer to eq. (66) with potential $\tilde{v}(r)$ $(v(r))$.

The new potential $\tilde{v}(r)$ is given in terms of K through [17]

$$\tilde{v}(r) - v(r) = -\frac{2}{r}\frac{d}{dr}\left[\frac{K(r,r)}{r}\right] \tag{82}$$

The inverse procedure summarizes as follows : once the potential and its wave functions are known (v can be identically zero, for example), one should give the spectral data. For the present case

$$S.D. = \{ \tilde{D}(\tau), \tau \geq 0 ; \tilde{\lambda}_K, \tilde{C}_K , K = 1,2,\ldots,\tilde{M} \} \tag{83}$$

where $\tilde{D}(\tau) \geq 0$ for $\tau > 0$ and $\tilde{D}(\tau) - 1 = O(\tau^{-2})$ for $\tau \to \infty$ (84)

$\tilde{\lambda}_K, \tilde{C}_K > 0$, $K = 1,\ldots,\tilde{M}$. [17]

This information allows to compute $\Omega(r,r')$ from eq. (81). Then from the GLM eq.(80) one gets $K(r,r')$ and finally $\tilde{v}(r)$ is obtained from eq.(82). Eq.(84) guarantees that eq.(80) is Fredholm type and hence the existence and uniqueness of the solutions $K(r,r')$. Moreover the wave functions corresponding to the potential $\tilde{v}(r)$ are given by

$$\tilde{\varphi}(\lambda,r) = \varphi(\lambda,r) + \int_r^\infty \frac{ds}{s^2} K(r,s)\, \varphi(\lambda,s) \tag{84'}$$

When the continuum part of the SD is trivial ($D(\tau) = 1$, $\tau \in \mathbb{R}^+$) the Gelfand-Levitan equation admits a closed solution because the kernel $\Omega(r,r')$ becomes degenerate. We find

$$K(r,r') = -\frac{2}{\pi}\sqrt{rr'} \sum_{1 \leq i,j \leq M} c_i\, K_{\lambda_i}(r) [W^{-1}(r)]_{ij} K_{\lambda_j}(r') \tag{85}$$

where $W(r)$ is a M by M matrix with entries

$$W_{ij}(r) = \delta_{ij} + \frac{2}{\pi} c_i \int_r^\infty \frac{ds}{s} K_{\lambda_i}(s)\, K_{\lambda_j}(s) \tag{86}$$

The corresponding potential results

$$v(r) = -2\, \nabla^2 \operatorname{Log} \det W(r) \tag{87}$$

where $\nabla^2 \equiv \frac{d^2}{dr^2} + \frac{1}{r}\frac{d}{dr}$. The solution of the Schrödinger equation for this potential and any λ follows directly by replacing $K(r,r')$ given by eq.(85) into (84').

The main point is that any potential $v(r)$ satisfying eq.(73) is in one to one correspondence with scattering data (83) under conditions (84) This is the IST in the angular momentum for the Schrödinger equation.

We shall now derive the trace identities associated with this IST.

The starting point is the relation between the Fredholm determinant of the operator

$$\hat{L} = \hat{L}_0 + r^2 v(r)$$

$$\hat{L}_0 = -r^2 \frac{d^2}{dr^2} + r^2 - \frac{1}{4} \tag{88}$$

and the Jost function. The Fredholm determinant is defined by

$$\Delta(\lambda^2) = \mathrm{Det}\left(\frac{\hat{L} + \lambda^2}{\hat{L}_0 + \lambda^2}\right) \tag{89}$$

Then,

$$\frac{\partial}{\partial \lambda^2} \operatorname{Log} \Delta(\lambda^2) = \int_0^\infty \frac{dr}{r^2} \left[G(r,r;\lambda) - G_0(r,r;\lambda) \right] \tag{90}$$

where

$$(\hat{L} + \lambda^2)\, G(r,r';\lambda) = r^2 \delta(r-r')$$

$$(\hat{L}_0 + \lambda^2)\, G_0(r,r';\lambda) = r^2 \delta(r-r')$$

These Green's functions can be expressed as

$$G(r,r';\lambda) = \frac{f(r_<,\lambda)\, \varphi(r_>,\lambda)}{2F(\lambda)} \tag{91}$$

$$G_0(r,r';\lambda) = \sqrt{rr'}\; I_\lambda(r_<)\, K_\lambda(r_>) \tag{92}$$

Now, the integral in eq.(90) can be computed with the aid of Wronskian relations.

$$\frac{\partial}{\partial\lambda^2} \operatorname{Log} \Delta(\lambda^2) = \lim_{\varepsilon\to 0^+} \lim_{\mu\to\lambda} \int_\varepsilon^\infty \frac{dr}{r^2} \left[\frac{f(\lambda,r)\,\varphi(\mu,r)}{2F(\lambda)} \right.$$

$$\left. - \frac{f_0(\lambda,r)\,\varphi_0(\mu,r)}{2F_0(\lambda)} \right] = \lim_{\varepsilon\to 0^+} \lim_{\mu\to\lambda} \frac{\varepsilon^{\lambda-\mu}}{2\mu(\lambda-\mu)} \left[\right.$$

$$\left. \frac{F_0(\mu)}{F_0(\lambda)} - \frac{F(\mu)}{F(\lambda)} \right] = \frac{d}{d\lambda^2} \operatorname{Log} \frac{F(\lambda)}{F_0(\lambda)}$$

Noting that both $\Delta(\lambda^2)$ and $F(\lambda)/F_0(\lambda)$ tend to unity for $\lambda\to\infty$ we get

$$\Delta(\lambda^2) = F(\lambda)/F_0(\lambda) \tag{93}$$

This ratio is an analytic function of λ for $\operatorname{Re}\lambda > 0$ and it equals one at infinity. Hence

$$\operatorname{Log} \Delta(\lambda^2) = \sum_{m=1}^{\infty} C_m \lambda^{-m} \tag{94}$$

We shall compute the C_m in two ways : firstly in terms of $v(r)$ and secondly from the Jost function in terms of the S.D.

We define the functions

$$\psi(\lambda,r) = \operatorname{Log} f(\lambda,r)/f_0(\lambda,r) \tag{95}$$

From eqs.(71) and (75) follow that

$$\psi(\lambda,\infty) = \operatorname{Log} \frac{F(\lambda)}{F_0(\lambda)} \quad ; \quad \psi(\lambda,0) = 0 \tag{96}$$

Hence

$$\mathrm{Log}\,\frac{F(\lambda)}{F_0(\lambda)} = \int_0^\infty dr\, \frac{\partial \Psi(\lambda,r)}{\partial r} \tag{97}$$

$\Psi(\lambda,r)$ satisfies the Ricatti equation

$$\Psi'' + \Psi'^2 + \left[\frac{1}{r} + 2\frac{d}{dr}\log I_\lambda(r)\right]\Psi' = v(r) \tag{98}$$

$\Psi(\lambda,r)$ can be expanded in powers of λ^{-1}

$$\Psi' \equiv \frac{\partial \Psi}{\partial r} = \sum_{n=1}^{\infty} h_n(r)\,\lambda^{-n} \tag{99}$$

Hence from eqs. (94) and (97)

$$C_n = \int_0^\infty dr\, h_n(r) \tag{100}$$

and from eq. (98)

$$h_1(r) = r\,v(r)/2$$

$$h_{n+1}(r) = -\frac{r}{2}\left\{h_n'(r) + \sum_{m=1}^{n}\left[h_{n-m}(r) + P_{n-m}(r)\right]h_m(r)\right\}$$

$$n \geq 1 \tag{101}$$

Here $h_0(r) \equiv 0$ and the $P_m(r)$ are polynomials of degree less or equal to $2m-1$ defined by

$$\sum_{m=1}^{\infty}\frac{P_m(r)}{\lambda^m} = 2\frac{d}{dr}\mathrm{Log}\left[r^{-\lambda} I_\lambda(r)\right] \tag{102}$$

$$P_0(r) = 1/r \quad , \quad P_1(r) = r \quad , \quad P_2(r) = -r \; , \; \ldots\ldots$$

One gets in this way that the $h_{2n}(r)$ are total derivatives and hence $C_{2n} \equiv 0$. The first terms of the expansion of $\mathrm{Log}\,\Delta(\lambda^2)$ read

$$\operatorname{Log}\Delta(\lambda^2) = \frac{1}{2\lambda}\int_0^\infty r v(r)\,dr - \frac{1}{8\lambda^3}\left(1+\frac{1}{\lambda^2}\right)\int_0^\infty r^3\,dr\left[v^2+2v\right]$$

$$+\frac{1}{16\lambda^5}\int_0^\infty r^5\,dr\left[\frac{1}{2}\left(\frac{dv}{dr}\right)^2+(v+1)^3-1\right]+O(\lambda^{-7})$$

$$\lambda\to\infty \tag{103}$$

Now we turn to express $\log\Delta(\lambda^2)$ in terms of SD.

A dispersion for $F(\lambda)$ can be derived by considering the logarithm of the function

$$V(\lambda) = \prod_{K=1}^{n}\frac{\lambda+\lambda_K}{\lambda-\lambda_K}\,\frac{F(\lambda)}{F_0(\lambda)}$$

This function is analytic and non-zero in the right half-plane and $V(\infty)=1$. From a Cauchy type integral of $\operatorname{Log}V(\lambda)$ and eq. (78) it follows that

$$\frac{F(\lambda)}{F_0(\lambda)} = \prod_{K=1}^{n}\frac{\lambda-\lambda_K}{\lambda+\lambda_K}\exp\left[\frac{2\lambda}{\pi}\int_0^\infty\frac{d\tau}{\tau^2+\lambda^2}\ln D(\tau)\right] \tag{104}$$

and

$$\frac{F(\lambda)}{F_0(\lambda)} = \prod_{K=1}^{M}\left(1-\frac{\lambda_K^2}{\lambda^2}\right)\exp\left[-\frac{2}{\pi}\int_0^\infty\frac{\tau\,d\tau}{\tau^2+\lambda^2}\,\delta(\tau)\right] \tag{105}$$

The asymptotic expansion of $\log\Delta(\lambda^2)$ in terms of SD follows starting from the identity

$$\oint_{C}\frac{d\lambda}{2\pi i}\,\lambda^{2z}\,\frac{d}{d\lambda}\operatorname{Log}\frac{F(\lambda)}{F_0(\lambda)} = -\sum_{K=1}^{M}\lambda_K^{2z} \tag{106}$$

where $0<\operatorname{Re}z<1/2$ and C is a contour from $-i\infty$ to $+i\infty$ closed by a large semicircle on the right-half plane. We can express the l.h.s. of eq. (106) in terms of $D(\tau)$ and $\delta(\tau)$ with the result

$$\cos \pi z \int_0^\infty d\tau \, \tau^{2z-1} \, \delta(\tau) + \sin \pi z \int_0^\infty d\tau \, \tau^{2z-1} \ln D(\tau) = -\frac{\pi}{2z} \sum_{K=1}^{M} \lambda_K^{2z} \tag{107}$$

We obtain by analytic continuation up to integer and half integer values of z

$$\sum_{K=1}^{M} \lambda_K^{2m} = \frac{m \, (-1)^{m+1}}{(2m-1)!} \, \frac{2}{\pi} \int_0^\infty d\tau \left(\tau \frac{d}{d\tau} \tau \right)^{2m-1} \delta(\tau) \tag{108}$$

$$\sum_{K=1}^{M} \lambda_K^{2m-1} = -\left(m - \frac{1}{2}\right) C_{2m-1} + \frac{(m-1/2)\,(-1)^{m+1}}{(2m-2)!} \, \frac{2}{\pi} \int_0^\infty d\tau \left(\tau \frac{d}{d\tau} \tau \right)^{2m-2} \ln D(\tau) \tag{109}$$

where $m = 1, 2, \ldots$ and we have used eq. (94) and ref [20].

The residue at $z = 0$ of eq. (107) gives Levinson's theorem

$$\delta(0+) = \pi M \tag{110}$$

The trace identities follow by expressing the C_{2m-1} in eqs. (108-109) in terms of the potential through eq. (103). We get for the first identities

$$\int_0^\infty r \, dr \, v(r) = -4 \sum_K \lambda_K + \frac{4}{\pi} \int_0^\infty d\tau \ln D(\tau) \tag{111}$$

$$0 = \sum_K \lambda_K^2 + \frac{2}{\pi} \int_0^\infty \left(\tau \frac{d}{d\tau} \tau \right) \delta(\tau) \tag{112}$$

$$\int_0^\infty r^3 dr \left[v(r)^2 + 2v(r) \right] = \frac{16}{3} \sum_K \lambda_K^3 + \frac{8}{\pi} \int_0^\infty d\tau \left(\tau \frac{d}{d\tau} \tau \right)^2 \ln D(\tau) \tag{113}$$

$$0 = \sum_{K} \lambda_K^4 + \frac{1}{3}\frac{2}{\pi}\int_0^\infty d\tau \left(\tau \frac{d}{d\tau}\tau\right)^3 \delta(\tau) \tag{114}$$

$$\int_0^\infty r^5 dr \left\{ \frac{1}{2}\left(\frac{dv}{dr}\right)^2 + [v(r)+1]^3 - 1 \right\} =$$

$$= 32\sum_K \left(\frac{\lambda_K^3}{3} - \frac{\lambda_K^5}{5}\right) + \frac{16}{\pi}\int_0^\infty d\tau \left[\left(\tau\frac{d}{d\tau}\tau\right)^2 + \left(\tau\frac{d}{d\tau}\tau\right)^4 \frac{1}{2}\right] \ln D(\tau) \tag{115}$$

$\mathcal{C}_{2m-1}$ can be expressed for any m as an integral of r^{2m-1} times a local polynomial in $v(r)$ and its derivatives. Hence, the trace identity for $\mathcal{C}_{2m-1}$ express an integral of a rotationally invariant field configuration over a $2m$ – dimensional space in terms of SD.

It must be noted that the derivations of the nth. trace identity assumes

$$\int_0^\infty r^m \, |v(r)| \, dr < \infty \tag{116}$$

Otherwise r^m must be interpreted in principal value at $r = \infty$

$$r^m \longrightarrow \frac{(-1)^m}{m!}\left(r\frac{d}{dr}r\right)^m$$

VI - THE IST IN THE ANGULAR MOMENTUM APPLIED TO THE NON-LINEAR SIGMA MODEL

The classical non linear sigma models are discussed in extenso in Eichenherr's lectures . They can be quantized in the perturbative expansion on the coupling constant or in the $1/N$ expansion [18].

The generating functional of the $O(N)$ non-linear σ– model can be written as

$$Z[J] = \int\!\!\int \mathcal{D}\vec{\sigma} \prod_x \delta(\vec{\sigma}(x)^2 - N) \exp\left\{-\frac{1}{g_0^2}\int dx \left[\partial_\mu \vec{\sigma}^2 + \vec{J}\cdot\vec{\sigma}\right]\right\} \tag{117}$$

where $\vec{\sigma} = (\sigma_1, \dots, \sigma_N) \in S^N$ and g_0 stands for the bare coupling constant. We work in two dimensional euclidean space. In order to generate systematically the $1/N$ expansion, it is convenient to use the Fourier representation of the functional Dirac delta and then to integrate over the $\vec{\sigma}$ field. One gets in this way

$$Z[J] = \int \mathcal{D}\alpha \, \exp\Big\{ -\frac{N}{2} \operatorname{Log} \operatorname{Det}\left[-\partial^2 - i g_0^2 \alpha(\cdot)\right] \tag{118}$$
$$- i N \int \alpha \, dx + \frac{g_0^2}{4} \int dx \, dy \; G(x,y) \; \vec{J}(x) \cdot \vec{J}(y) \Big\}$$

This functional integral has a constant saddle point at $\alpha(x) = \alpha_c$ where

$$iN \left[\frac{g_0^2}{2} (x| \frac{1}{-\partial^2 - i g_0^2 \alpha_c} |x) - 1 \right] = 0 \tag{119}$$

where the relation

$$\frac{\delta}{\delta\alpha(x)} \operatorname{Log} \det M = \operatorname{Tr}\left(M^{-1} \frac{\delta M}{\delta \alpha(x)} \right) \tag{120}$$

has been used. In two dimensions

$$(x| \frac{1}{-\partial^2 + m^2} |x) = \int^{\Lambda} \frac{d^2 p}{(2\pi)^2} \frac{1}{p^2 + m^2} = \frac{1}{2\pi} \ln\left(\frac{\Lambda^2}{m^2} + 1\right) \tag{121}$$

where Λ is a momentum cut off. Hence from eqs. (119) and (121)

$$g_0^2 = \frac{1}{8\pi \ln \frac{\Lambda^2}{m^2}} \;, \qquad \alpha_c = i m^2 / g_0^2 \tag{122}$$

Here m^2 turns to be the renormalized mass. The particle spectrum of the theory is constituted by N mesons with mass m in the adjoint representation of $O(N)$.

Eq. (121) exhibits the asymptotic freedom of the two dimensional non-linear sigma model.

As usual one must shift the integration variable

$$\alpha(x) = \alpha_c + i g_0^{-2} v(x)$$

Here $v(x)$ is the new integration variable and we assume $v(\infty) = 0$. The functional integral results

$$Z = \int \mathcal{D}v \; \exp - \frac{N}{2} S_{eff}[v] \tag{123}$$

$$S_{eff}[v(.)] = \mathrm{Log\,Det}\left(\frac{-\partial^2 + m^2 + v(.)}{-\partial^2 + m^2}\right) - \frac{2}{g_0^2}\int d^2x \; v(x) \tag{124}$$

and the $1/N$ series follows by expanding here around $v = 0$.

The IST transformation can be applied to S_{eff} only if v depends on one variable. We assume that v is rotationally invariant : $v = v(r)$. This assumption is not too restrictive because the lowest extended solutions has usually a maximum of symmetry.

We can expand now S_{eff} in partial waves. This expansion of S_{eff} with a momentum cut-off is rather subtle. We recall that the counterterm $-2 g_0^{-2} \int dx \, v(x)$ substracts from the functional determinant its linear part in $v(x)$. In partial waves this prescription gives, using eqs.(92) and (120)

$$S_{eff}[v] = \sum_{\ell=-\infty}^{+\infty} \left[\ln \Delta(\ell^2) - \int_0^\infty r \, v(r) \, I_\ell(r) \, K_\ell(r) dr \right] \tag{125}$$

where we choose units such that $m = 1$.

By using addition theorems one can perform the sum over ℓ of the cylindric functions.

$$S_{eff} = 2 \sum_{\ell=1}^{\infty} \left[\log \Delta(\ell^2) - \frac{1}{2\ell} \int_0^{\infty} r\, v(r)\, dr \right] +$$

$$+ \log \Delta(0) + \int_0^{\infty} dr\; v(r)\; r\, \ln(r/r_0) \; . \tag{126}$$

Here r_0 is a numerical constant. This expression can also be derived by dimensional regularization.

Let us now express S_{eff} in terms of SD. Except for the last term in eq.(126) we know how to do that from sec. IV.

In order to express

$$Q[v] = \int_0^{\infty} r\, dr\; \ln r\; v(r) \tag{127}$$

in terms of SD we compute $\delta v(r)$ for small variations of the SD from the Gelfand-Levitan eq.(80). $Q[v]$ is a typical renormalization effect appearing for the first time here because of the UV divergences

For small variations of the SD we have from eqs.(80) and (82)

$$\delta v(r) = -\frac{2}{r}\frac{d}{dr}\,\delta K(r,r) = \frac{2}{r}\frac{d}{dr}\,\delta\Omega(r,r)$$

We get in this way

$$\frac{\delta v(r)}{\delta D(\tau)} = -\frac{8}{r}\,\frac{\tau\, sh\, \pi\tau}{D(\tau)^3}\,\frac{d}{dr}\,\varphi(r,i\tau)^2 \tag{128}$$

and similarly for $\partial v/\partial\lambda_K$ and $\partial v/\partial c_K$. From these relations we compute the functional derivatives of Q and they result expressed in closed form in terms of the SD. Finally an explicit expression for Q follows

$$Q = \frac{4}{\pi^2} \fint_0^\infty \fint_0^\infty \frac{\tau^2 + \tau'^2}{(\tau^2 - \tau'^2)^2} \ln D(\tau) \ln D(\tau') \, d\tau \, d\tau' +$$

$$+ \frac{4}{\pi} \int_0^\infty d\tau \ln D(\tau) \left[\operatorname{Re} \Psi(i\tau) + \ln 2 - \sum_{K=1}^{M} \frac{2\lambda_K}{\tau^2 + \lambda_K^2} \right] +$$

$$+ 2 \sum_{K=1}^{M} \ln \left[\frac{\pi \lambda_K}{c_K \, \Gamma(\lambda_K)^2} \, 2^{2(1-\lambda_K)} \right] + \sum_{1 \leq K \neq L \leq M} \log \left(\frac{\lambda_K + \lambda_L}{\lambda_K - \lambda_L} \right)^2 . \tag{129}$$

Here $\Psi(z) = \frac{d}{dz} \operatorname{Log} \Gamma(z)$

This relation resembles to a trace identity but it mixes different SD and it contains the normalization coefficients C_K . $Q[v]$ appears as the renormalized trace of the solution of the Gelfand-Levitan equation.

We can now express S_{eff} in terms of the scattering data using eqs. (126), (93), (104), (111), (127) and (129).

$$S_{eff}[v] = \frac{4}{\pi^2} \fint_0^\infty \fint_0^\infty \frac{\tau^2 + \tau'^2}{(\tau^2 - \tau'^2)^2} \ln D(\tau) \ln D(\tau') \, d\tau \, d\tau' +$$

$$+ \ln D(0) + i\pi M - \frac{8}{\pi} \int_0^\infty \sum_{K=1}^{M} \frac{\lambda_K \ln D(\tau)}{\lambda_K^2 + \tau^2} \, d\tau +$$

$$+ 2 \sum_{K=1}^{M} \ln \left[\frac{4\lambda_K^2}{c_K} \sin \pi \lambda_K \right] + \sum_{1 \leq K \neq L \leq M} \ln \left(\frac{\lambda_K + \lambda_L}{\lambda_K - \lambda_L} \right)^2 . \tag{130}$$

It is remarkable that S_{eff} admits an explicit expressions in terms of the SD. This is also the case for the N- dimensional anharmonic oscillator [see II]. However, here the action does not separate. We have here couplings between the SD and the normalization coefficients are present. The coupling and the c_K come from $Q[v]$ which is a typical renormalization effect clearly absent in quantum mechanics.

The mapping between $v(r)$ and the SD is a local diffeomorphism. Hence we can find stationary points of S_{eff} from

$$0 = \frac{\delta S_{eff}}{\delta \ln D(\tau)} = \frac{4}{\pi} \frac{d}{d\tau} \delta(\tau) \quad ; \quad \tau > 0 \tag{131}$$

$$0 = \frac{\partial S_{eff}}{\partial c_m} = - \frac{2}{c_m} \quad ; \quad m = 1, 2, \ldots, M .$$

Where we have used eqs.(104) and (105). Because $\delta(\infty) = 0$ the solution of eq. (131) is

$$\delta(\tau) = 0 \quad , \quad \tau > 0 \tag{132}$$

Eq. (131) holds only if $M = 0$ because $c_m = \infty$ implies a vanishing eigenfunction. Then, in absence of bound states the equation $\partial S_{eff} / \partial \lambda_m = 0$ does not need to be considered. From eq.(132) and $M = 0$ we get $D(\tau) = 1$ hence the only extremum of S_{eff} is $v(r) = 0$.

The absence of non-trivial solutions is associated to unstability under a unlimited expansion of a given configuration $v(r)$ [4].

This is like Derrick's theorem in renormalizable scalar models where non trivial configuration are unstable but under collapse. This tendency to expansion is associated to the ultraviolet freedom of the non-linear σ- model. Unstability under unlimited expansion relates to ultraviolet stability (asymptotic freedom).

Although our proof restricts to rotationally invariant configurations the unboudness of S_{eff} from above and below is true in general and very probably excludes also the existence of non-symmetric solutions.

If S_{eff} would have a saddle point with finite action, say S_0, it will determine the large orders of the $1/N$ expansion as it is the case for the anharmonic oscillator (Sec. II). The K th. order of the series will behave like

$$K! \; S_0^{-K} \qquad (K \gg 1)$$

The exact S- matrix of the non-linear σ- model is known [19]. If one expands it in powers of $1/N$, one finds for the coefficient of N^{-K} for large K

$$\frac{1}{K}\left(\frac{2\pi}{i\theta}\right)^K \qquad \text{and} \qquad \frac{1}{K}\left(\frac{2\pi}{i\theta+\pi}\right)^K$$

where θ is the rapidity. Hence the serie converges for

$$\frac{1}{N} < \min\left[\frac{|\theta|}{2\pi}, \frac{|\theta - i\pi|}{2\pi}\right]$$

We see that this observed behavior is consistent with a saddle point computation only if $S_0 = \infty$ which is precisely our limiting case. All that suggests us to conjecture that the $1/N$ expansion of the non-linear σ- model is convergent not only for the S- matrix but also for the Green functions.

The graphs in the $1/N$ series of the sigma model look at first sight like those of any renormalizable theory. The fact that they give a convergent series on-shell at two dimensions shows that tremendous cancellations take place. The absence of saddle points suggests that these cancellations can also take place off-shell.

REFERENCES

[1] See for reviews

[1a] A.C. Scott, F.Y.F. Chu and D.W. Mc Laughlin, Proc. IEEE 61, 1443 (1973).

[1b] L.D. Faddeev, Journal of Math. Phys. 4, 72 (1973) and Journal of Soviet Mathematics 5, 334 (1976).

[1c] M. Ablowitz, D.J. Kaup, A.C. Newell and H. Segur, Stud. Appl. Math. 53, 249 (1974).

[1d] K. Chadan and P.C. Sabatier, Inverse Problems in Quantum Scattering Theory, Springer Verlag, New York 1977.

[2] H.J. de Vega, Commun. Math. Phys. 70, 29 (1979).

[3] H.J. de Vega, Phys. Rev. D 21, 395 (1980) - See also Ref.[5]

[4] H.J. de Vega, Phys. Lett. 98 B, 280 (1981).

[5] H.J. de Vega, Phys. Rev. D 22, 2400 (1980).

[6] H.J. de Vega, (in preparation).

[7] R.L. Stratonovich, Sov. Phys. Dokl. 2, 416 (1957).

[8] E. Brézin, J.C. Le Guillou and J. Zinn-Justin, in Field Theoretical Approach to Critical Phenomena, edited by C. Domb and M.S. Green, Academic Press, New York 1976, Vol. 6.

[9] The computation of large orders in perturbation theory from instantons can be found in
L.N. Lipatov, J.E.T.P. Letters 25, 105 (1977), J.E.T.P. 45, 216 (1977)
E. Brézin, J.C. Le Guillou and J. Zinn-Justin, Phys. Rev. D 15, 1544 and 1558 (1977).

[10] S. Hikami and E. Brézin, J. Phys. A 12, 759 (1979).

[11] L.D. Faddeev and V.E. Zakharov, Funct. Anal. and Appl. 5, 280 (1971)
For a review see
L.D. Faddeev in "Solitons" page 339, Ed. by R.K. Bullough and P.J. Caudrey, Springer Verlag, Berlin (1980).

[12] J.L. Gervais and A. Jevicki, Nucl. Phys. B 110, 93 (1976).

[13] T. Banks, C.M. Bender and T.T. Wu, Phys. Rev. D 8, 3346 and 3366 (1973).

[14] J. Zinn-Justin, Saclay preprint, DPh T/160 (1979).

[15] R.J. Cant, Phys. Lett. 95 B, 380 (1980).

[16] See for example
H.J. de Vega. J.L. Gervais and B. Sakita, Nucl. Phys. B 139, 20 (1978) and Ref. 14.

[17] H.J. de Vega, LPTHE preprint PAR 80/29 (1980). To be published in Comm. Math. Phys.

[18] E. Brézin and J. Zinn-Justin, Phys. Rev. Lett. 36, 691 (1976),
Phys. Rev. B 14, 3110 (1976).
W. Bardeen, B. Lee and R. Shrock, Phys. Rev. D 14, 985 (1976).

[19] A.B. Zamolodchikov and Al. B. Zamolodchikov, Ann. Phys. 80, 253 (1979).

[20] I.M. Gelfand and G.E. Shilov, Generalized functions, Academic Press, New York 1964.

CLASSICAL SOLUTIONS TO CP^{n-1} MODELS AND THEIR GENERALIZATIONS

W.J. Zakrzewski
Dept. of Mathematics,
University of Durham
Durham, U.K.

1. Introduction

It is generally believed that nonabelian gauge theories are likely to play an important role in any field theoretical description of the theory of elementary particles. These theories, in case of an SU(2) theory are defined in terms of a Lagrangian density

$$L = \mathrm{tr} F_{\mu\nu} \quad F_{\mu\nu} \qquad 1.1$$

where

$$F_{\mu\nu} = \partial_\mu A_\nu - \partial_\nu A_\mu + [A_\mu, A_\nu] \qquad 1.2$$

and where A_μ is an SU(2) valued vector function of a Euclidean 4-dimensional space-time.

One of the main stumbling blocks in making any progress with these theories is our lack of understanding how to perform functional integrations

$$\int DA_\mu \; e^{-\int d^4x \, L(A_\mu)} \; O(A_\mu) \qquad 1.3$$

in terms of which most quantities of the theory are given.

The only viable line of approach to calculate integrals like 1.3 that is available at present is the expansion around stationary points of the action and then perturbation theory of the resultant effective theory.

Thus one has to determine first the stationary points of the action. They are given by the Euler-Lagrange equations of the theory which are

$$D_\mu F_{\mu\nu} = \partial_\mu F_{\mu\nu} - [A_\mu, F_{\mu\nu}] = 0. \qquad 1.4$$

When written in terms of A_μ these equations are highly nonlinear second order partial differential equations. As is well known, due to the Bianchi identity

$$D_\mu F^*_{\mu\nu} = 0 \qquad 1.5$$

where

$$F^*_{\mu\nu} = \frac{1}{2} \varepsilon_{\mu\nu\alpha\beta} F_{\alpha\beta} \qquad 1.6$$

a subclass of solutions of equations 1.4 is provided by the solutions of the first order equations:

$$F_{\mu\nu} = \pm F^*_{\mu\nu} \qquad 1.7$$

This equation can be thought of as coming from

$$L = \pm Q \qquad 1.8$$

where $Q(x) = TrF_{\nu\mu}F^*_{\nu\mu}$ is the density of the topological charge. We are only interested in those solutions of equations of motion for which the action is finite - as it is only for these that we really know how to set up the perturbation theory of fluctuations around them.

All finite action solutions of equations 1.7 have been implicitly, though unfortunately not explicitly, determined by Atiyah, Hitchin, Drinfeld and Manin [1]. In the case of the plus(minus) sign in 1.7 - the corresponding finite action solutions are called instanstons (antiinstantons). A simple application of a Bogomolnyi bound [2] shows that the instanton and antiinstanton solutions correspond to the local minima of the action. Hence such solutions are stable under small fluctuations. Even though all finite action solutions were found it is not clear whether there are any further solutions of 1.4 which are of finite action and which are not solutions of 1.7. Had such noninstanton solutions existed they presumably would have also had to be included in the stationary point calculation of 1.3. Unfortunately, even though some progress in settling the question of their existence has been made this problem still awaits its solution.

Also the question of calculating the fluctuations about the instanton solutions has turned out to be a hard mathematical problem. Some progress has been made in the simplest possible case (O(A) = 1) [3] but it has become clear that further progress will not come too quickly. This is one of the reasons why in the last years people have turned their attention to simpler field theories which bear as much resemblance as possible to 4-dimensional nonabelian gauge theories.

Nonabelian gauge theories possess certain properties which are retained, although often in a modified form, by these simpler field theories. These properties are

1) conformal invariance of the action
2) existence of nontrivial solutions to the equations of motion
3) existence of a parameter N - to allow for a possible 1/N expan-

sion - in the case of SU(N) gauge theories

4) asymptotic freedom

and unproved but hoped for

5) confinement of fermions.

A model field theory, which possesses similar properties, but, bening a twodimensional theory, is hopefully simpler, was found by Golo & Perelomov and Eichenherr [4], and will be discussed in the next sections.

2. CP^{n-1} model and its instanton solutions

The CP^{n-1} model is based on the Lagrangian density

$$L = \overline{D_\mu Z_\alpha} \cdot D_\mu Z_\alpha \qquad \alpha = 1,..n \tag{2.1}$$

where

$$D_\mu Z_\alpha = [\partial_\mu - (\bar{Z}_\beta \partial_\mu Z_\alpha)]Z_\alpha \tag{2.2}$$

and where in addition the $Z_\alpha(x,y)$ field satisfies the constraint

$$\bar{Z}_\alpha \cdot Z_\alpha = 1 \tag{2.3}$$

and we consider only equivalence classes of fields which differ from each other by an overall space dependent phase factor

$$Z'_\alpha \simeq Z_\alpha \qquad \text{if } Z'_\alpha = Z_\alpha e^{i\phi(x,y)} \tag{2.4}$$

It can be shown that a theory based on this Lagrangian density to which a fermionic field contribution is added, as will be discussed later on, possesses all the properties mentioned in the previous section.

We are primarily interested in finding finite action solutions of the equations of motion.

It is easy to find the Euler-Lagrange equations for the Lagrangian density 2.1. Taken together with the constraint 2.3 they are

$$D_\mu D_\mu Z_\alpha + (\overline{D_\mu Z_\beta} \cdot D_\mu Z_\beta)\ Z_\alpha = 0 \tag{2.5}$$

and they are the CP^{n-1} analogue of the equations 1.4.

Like in the gauge theory case one can introduce a topological charge density

$$Q = i\ \varepsilon_{\mu\nu}\ \partial_\mu(\bar{Z}_\alpha \partial_\nu Z_\alpha) \tag{2.6}$$

and then consider equations coming from the relation 1.8

$$L = \pm Q$$

They are given by

$$D_\mu Z_\alpha = \pm i\varepsilon_{\mu\nu} D_\nu Z_\alpha \qquad 2.7$$

and they correspond to the equations 1.7. Their finite action solutions are again called instantons and antiinstantons respectively. To obtain a better insight into the CP^{n-1} model it is convenient to change the Euclidean variables (x,y) to the holomorphic and antiholomorphic variables

$$x_+ = x + iy$$
$$x_- = x - iy \qquad 2.8$$

and then rewrite all equations in terms these variables. We find that the action and the topological charge densities are given by

$$L = 2(|D_+Z_\alpha|^2 + |D_-Z_\alpha|^2)$$
$$Q = 2(|D_+Z_\alpha|^2 - |D_-Z_\alpha|^2) \qquad 2.9$$

where the derivatives denote the covariant derivative 2.2 in which the differentiation is performed with respect to the $x_\pm$ variables. In terms of the $x_\pm$ variables the first order equations 2.7 are just simply

$$D_+Z_\alpha = 0$$
$$D_-Z_\alpha = 0 \qquad 2.10$$

and as was shown in the original paper of d'Adda et al. [5] their solutions are given by

$$Z_\alpha = \frac{f_\alpha}{|f|} \qquad 2.11$$

where in the instanton (antiinstanton) case f_α is a function of only x_+ (x_-). The finiteness of the action imposes conditions on the components of f_α - they have to be rational functions of their argument.

However due to the invariance under an overall factor multiplication (due to 2.11 and 2.4) - it is sufficient to consider only polynomial components of f_α (with no overall factors). This was found already in the original paper of d'Adda [5] - in which a detailed discussion of the instanton (and antiinstanton) solutions was given.

We see that in contradistinction to the gauge theory case the form of all solutions to the first order differential equations (2.10) is very simple and explicit and this suggests that it may be not too difficult to find finite action solutions of equation 2.5 which are not solutions of 2.7.

Such a construction of all finite action solutions of equation 2.5

will be given in the next section [6]; their properties and also generalizations to other models will be discussed in the following sections [7,8,9]. This construction arose out of a work of Borchers and Garber [10,11] who have considered a similar problem in the case of the O(N) σ models. This work, with several modifications, could be adopted to the CP^{n-1} case, where it allows for an elegant mathematical pattern and brings out the geometry of the problem.

We finish this section by rewriting the equations of motion 2.5 in terms of D_+ and D_- variables. We obtain

$$D_+D_-Z_\alpha + D_-D_+Z_\alpha + \frac{1}{2} L Z_\alpha = 0 \qquad 2.12$$

However as

$$[D_+, D_-] = |D_+Z|^2 - |D_-Z|^2 = \frac{1}{2} Q \qquad 2.13$$

the equations of motion can be rewritten as

$$D_-D_+Z_\alpha + A_+Z_\alpha = 0 \qquad 2.14$$

or

$$D_+D_-Z_\alpha + A_-Z_\alpha = 0 \qquad 2.15$$

where

$$A_\pm = \frac{1}{4}(L \pm Q) = |D_\pm Z_\alpha|^2$$

It is the equation 2.15 that we shall solve in the next section.

3. General solutions

To construct general solutions of the equations of motion we first study consequences of their existence. Assume that $Z_\alpha(x_+, X_-)$ is such a finite action solution of the equations of motion (2.5). We shall show that

$$A^m_{ij} \equiv \overline{D^i_- Z_\alpha} \cdot D^j_+ Z_\alpha = 0 \quad \text{for } m \equiv i+j \geq 1 \qquad 3.1$$

It is clear that $A^1_{01} = A^1_{10} = 0$. Moreover, as was shown in ref. 12, the finiteness of the action and the conservation of energy momentum give $A^2_{1,1} = 0$. In a more general case we shall prove 3.1 by induction. Thus we assume that $A^m_{i,j} = 0$ for $i \leq i_o$, $j \leq j_o$, $m < m_o$. Then as

$$\partial_+(\bar{a}_\alpha \cdot b_\alpha) = (\overline{D_- a_\alpha} \cdot b_\alpha) + (\bar{a}_\alpha D_+ b_\alpha) \qquad 3.2$$

we see that

$$A^{m+1}_{i+1\ j} = \overline{D^{i+1}_{-} Z_\alpha}\ \ D^{j}_{+} Z_\alpha = \partial_+ (\overline{D^{i}_{-} Z_\alpha} \cdot D^{j}_{+} Z_\alpha) - \overline{D^{i}_{-} Z_\alpha}\ \ D^{j+1}_{+} Z_\alpha = -A^{m+1}_{i\ j+1} \qquad 3.3$$

and so it is enough to prove that $A^{m+1}_{0\ m+1} = 0$. First we show that $A^{m+1}_{0\ m+1} = 0$ is analytic, i.e. that it satisfies $\partial_- A^{m+1}_{0\ m+1} = 0$. We write

$$\partial_- A^{m+1}_{0\ m+1} = \overline{D_+ Z_\alpha} \cdot D^{m+1}_{+} Z_\alpha + \overline{Z}_\alpha \cdot D_- D^{m+1}_{+} Z_\alpha \qquad 3.4$$

and then observe that the first term can be written as

$$\overline{D_+ Z_\alpha} \cdot D^{m+1}_{+} Z_\alpha = \partial^{m}_{+} A_+ \qquad 3.5$$

This last result is shown by induction - it is clearly correct for m=0, and in general

$$\overline{D_+ Z_\alpha} D^{m+1}_{+} Z_\alpha = \partial_+ (\overline{D_+ Z_\alpha} \cdot D^{m}_{+} Z_\alpha) - \overline{D_- D_+ Z_\alpha}\ \ D^{m}_{+} Z_\alpha =$$
$$\partial_+ (\partial^{m-1}_{+} A_+) + A_+ \overline{Z}_\alpha \cdot D^{m}_{+} Z_\alpha = \partial^{m}_{+} A_+ \qquad 3.6$$

For the second term we have

$$(\overline{Z}_\alpha \cdot D_- D^{m+1}_{+} Z_\alpha) = \overline{Z}_\alpha (D_+ D_- - \tfrac{1}{2} Q) D^{m}_{+} Z_\alpha = \overline{Z}_\alpha D_+ D_- D^{m}_{+} Z_\alpha =$$
$$= \overline{Z}_\alpha \cdot D^{m}_{+} D_- Z_\alpha = -\overline{Z}_\alpha \cdot D^{m}_{+} A_+ Z_\alpha = - \partial^{m}_{+} A_+ \qquad 3.7$$

Hence $A^{m+1}_{0\ m+1}$ is analytic.

To show that $A^{m}_{0\ m}$ vanishes we use a variant of Liouville's theorem. As

$$|A^{m}_{0,m}|^2 \le |D^{m}_{+} Z_\alpha|^2 \qquad 3.8$$

we see that we have to show that $|D^{m}_{+} Z_\alpha|^2$ belongs to $\mathcal{L}^2(R^2)$. Of course from the finiteness of action we know that $|D_+ Z_\alpha| \in \mathcal{L}^2(R^2)$. To prove that $I_m = \int d^2x |D^{m}_{+} Z_\alpha|^2$ is finite one observes that the inversion transformed

$$Z'_\alpha(x_+, x_-) = Z_\alpha(x'_+, x'_-) \text{ where } x'_+ = \frac{1}{x_+} \text{ etc. ,}$$

is also a solution of the equations of motion and therefore real analytic. The proof is then completed repeating the discussion of ref. 10 and showing that I_m is the sum of integrals over the insides of the unit circles $|x_+| < 1$ and $|x'_+| < 1$ of real analytic functions and so is finite.

Thus we can now construct, for a given solution Z_α, two subspaces of C^n defined by

$$H_k = \{ D_-^i Z_\alpha, \quad i=1,2,\dots \}$$
$$H'_m = \{ D_+^i Z_\alpha, \quad i=1,2,\dots \} \qquad 3.9$$

where $k = \dim H_k$, $m = \dim H_m$.

From the previous discussion we know that those two subspaces are mutually orthogonal and are both orthogonal to Z. Clearly $k+m \leq n-1$. It can be shown that the case of $k+m < n-1$ corresponds to an embedding - so we shall only consider the case $k+m=n-1$. The spaces H_k and H'_m are spanned by the first k and m vectors in 3.9 respectively.

Next we shall find a holomorphic n component vector $f \epsilon \hat{H}_k = \{ Z_\alpha, H_k \}$ - i.e. $\partial_- f=0$, given in terms of Z_α. We define f_α by

$$\bar{f}_\alpha \cdot (D_-^i Z)_\alpha = \omega \delta^{ik} \qquad i = 0,\dots k \qquad 3.10$$

where

$$\partial_+ \omega - (\bar{Z}_\alpha \partial_+ Z_\alpha)\omega = 0. \qquad 3.11$$

To show that f_α is analytic we observe first that as $f_\alpha = \sum_{i=0}^{k} \mu_i D_-^i Z_\alpha$ $\partial_- f_\alpha \in \hat{H}_k$.

On the other hand for $i > 0$ we have

$$\overline{\partial_- f_\alpha} \cdot D_-^i Z_\alpha = \partial_+ (\bar{f}_\alpha \cdot D_-^i Z_\alpha) - \bar{f}_\alpha (\partial_+ D_-^i Z_\alpha) \qquad 3.12$$

Thus as $D_+ D_-^i Z$ is a linear combination of $Z_\alpha, \dots D_-^{i-1} Z_\alpha$ we see that 3.12 can be rewritten as

$$\overline{\partial_- f_\alpha} \cdot D_-^i Z_\alpha = \delta^{ik} [\partial_+ \omega - \bar{Z}_\alpha \partial_+ Z_\alpha \ \omega] = 0. \qquad 3.13$$

For $i=0$ we have $\bar{f}_\alpha Z_\alpha = 0$ which implies that $f_\alpha \in H_k$ and so $\partial_- f_\alpha \in H_K$ thus showing that $\partial_- f_\alpha Z_\alpha = 0$. Hence $\partial_- f_\alpha$ is orthogonal to H_k and so must vanish:

$$\partial_- f_\alpha = 0 \qquad 3.14$$

It is not difficult to show that f_α must be meromorphic - since it can be expressed as a rational function of Z_α and its derivatives which are real analytic; moreover as solutions to the equations of motion are invariant under the conformal transformation $x_+ \to x_+^{-1}$ we see that f_α must be rational.

Next we shall show that $\hat{H}_k$ is spanned by $f_\alpha, \partial_+ f_\alpha \dots \partial_+^k f_\alpha$ and that it is possible to express Z_α in terms of this basis. To this end we first show that

$$\overline{\partial_+^\ell f_\alpha} \cdot D_-^i Z_\alpha = (-1)^\ell \omega \ \delta^{i+\ell,k} \qquad 0 \leq i + \ell \leq k \qquad 3.15$$

For i = ℓ = 0 this is just $\bar{f}_\alpha \cdot Z_\alpha = 0$; using induction on i+ℓ we find

$$\partial_+^\ell f_\alpha \cdot D_-^i Z_\alpha = \partial_-(\overline{\partial_+^{\ell-1} f_\alpha} D_-^i Z_\alpha) - \overline{\partial_+^{\ell-1} f_\alpha} \cdot \partial_- D_-^i \, Z_\alpha = \tag{3.16}$$

$$= -\overline{\partial_+^{\ell-1} f_\alpha} D_-^{i+1} Z_\alpha = (-1)^\ell \bar{f}_\alpha D_-^{i+\ell} Z_\alpha = (-1)^\ell \omega \delta^{i+\ell,k}$$

Hence

$$\overline{\partial_+{}^\ell f_\alpha} \cdot Z_\alpha = (-1)^\ell \omega \delta^{\ell,k} \qquad \text{for } 0 \leq \ell \leq k \tag{3.17}$$

and so

$$\overline{\partial_+^\ell f_\alpha} \, D_+ Z_\alpha = \partial_+(\overline{\partial_+^\ell f_\alpha} \cdot Z_\alpha) - (\bar{Z}_\alpha \partial_+ Z_\alpha)(\overline{\partial_+^\ell f_\beta} Z_\beta) = (-1)^\ell \delta^{\ell,k}[\partial_+ \omega - (\bar{Z}_\alpha \partial_+ Z_\alpha)\omega] = 0$$

and by induction

$$\overline{\partial_+^\ell f_\alpha} \cdot D_+^i Z_\alpha = \partial_+(\overline{\partial_+^\ell f_\alpha} \cdot D_+^{i-1} Z_\alpha) - (\bar{Z}_\alpha \partial_+ Z_\alpha)(\overline{\partial_+^\ell f_\beta} \cdot D_+^{i-1} Z_\beta) = 0 \tag{3.18}$$

Thus we have shown that $\partial_+^\ell f_\alpha$ is orthogonal to H_m' and so that $\partial_+^\ell f_\alpha \in \hat{H}_k$ for ℓ = 0,1...k. As $f_\alpha, \partial_+ f_\alpha, .. \partial_+^k f_\alpha$ are all independent we see that $\hat{H}_k$ is spanned by them. We now proceed to express Z_α in terms of f_α, $\partial_+ f_\alpha, \ldots \partial_+^k f_\alpha$. First we define a matrix M by

$$M_{\ell i} = \overline{\partial_+^\ell f_\alpha} \cdot \partial_+^i f_\alpha \qquad i, \ell = 0, \ldots k-1 \tag{3.19}$$

and then define a vector $\hat{Z}_\alpha$ of $\hat{H}_k$ by

$$\hat{Z}_\alpha^{(k)} = (-1)^k [\partial_+^k f_\alpha - \sum_{i,\ell=0}^{k-1} M_{i\ell}^{-1} \partial_+ M_{\ell,k-1} \, \partial_+^i f_\alpha] \tag{3.20}$$

The vector $\hat{Z}_\alpha$ is orthogonal to $\hat{L}_k$ as

$$\overline{\partial_+^j f_\alpha} \cdot (\partial_+^k f_\alpha - \sum_{i,\ell=0}^{k-1} M_{i\ell}^{-1} \partial_+ M_{\ell,k-1} \partial_+^i f_\alpha) =$$

$$= \overline{\partial_+^j f_\alpha} \cdot \partial_+{}^k f_\alpha - \partial_+ M_{jk-1} = \overline{\partial_+^j f_\alpha} \cdot \partial_+{}^k f_\alpha - \partial_+(\overline{\partial_+^j f_\alpha} \cdot \partial_+^{k-1} f_\alpha) \tag{3.21}$$

$$= -\overline{\partial_- \partial_+^j f_\alpha} \cdot \partial_+^{k-1} f_\alpha = 0$$

where $0 \leq j \leq k-1$ and where $\hat{L}_k$ denotes a k dimensional subspace of $\hat{H}_k$. On the other hand Z_α is also orthogonal to that subspace as can be seen from 3.16 - hence it follows that Z_α and $\hat{Z}_\alpha$ are proportional to each other, and since Z_α is a unit vector defined up to a phase factor this means that

$$Z_\alpha = \frac{\hat{Z}_\alpha}{|\hat{Z}|} \tag{3.22}$$

Let us point out at this stage that the gauge freedom $Z_\alpha \to Z_\alpha e^{i\phi}$ implies an arbitrariness $\omega \to a(x_-) e^{i\phi} \omega$. Choosing $e^{i\varphi} = \frac{\bar{a}(x+)}{a(x_-)}$ we see that this corresponds to $\omega \to \bar{a}(x_+)\omega$ which in turn corresponds to $f_\alpha \to \bar{a}(x_+) f_\alpha$

which of course preserves analyticity. With the particular choice of gauge as in 3.22 we find

$$\overline{\partial_+^k f_\alpha} \cdot Z_\alpha = (-1)^k |\hat{Z}_\alpha| \qquad 3.23$$

and so

$$\omega = |\hat{Z}_\alpha| \qquad 3.24$$

which can be shown to fulfil 3.11.

It now remains to be shown that Z_α defined by 3.22 and 3.20 satisfies the equations of motion. To see this we observe that

$$\overline{\partial_+^i f_\alpha} \cdot Z_\alpha = (-1)^k |\hat{Z}| \delta^{ik} \qquad i=0,1..k \qquad 3.25$$

and that (since $\bar{\hat{Z}}_\alpha \partial_- \hat{Z}_\alpha = 0$)

$$D_- Z_\alpha = \frac{1}{|\hat{Z}|}(\partial_- \hat{Z}_\alpha - \frac{1}{|\hat{Z}|^2}(\bar{\hat{Z}}_\beta \partial_- \hat{Z}_\beta)\hat{Z}_\alpha) = \frac{1}{|\hat{Z}|}\partial_- \hat{Z}_\alpha \in \hat{H}_k \qquad 3.26$$

Then we have that

$$D_+ D_- Z_\alpha \in \hat{H}_k . \qquad 3.27$$

On the other hand

$$D_+ D_- Z_\alpha = \frac{1}{|\hat{Z}|} \partial_+ \partial_- \hat{Z}_\alpha - \frac{1}{|\hat{Z}|^3} (\bar{\hat{Z}}_\beta \partial_+ \hat{Z}_\beta) \partial_- \hat{Z}_\alpha \qquad 3.28$$

For $i = 0, \ldots k-1$

$$\overline{\partial_+^i f_\alpha} \partial_+ \partial_- \hat{Z}_\alpha = \partial_+ (\overline{\partial_+^i f_\alpha} \cdot \partial_- \hat{Z}_\alpha) = \partial_+ \partial_- (\overline{\partial_+^i f_\alpha} \hat{Z}_\alpha) - \partial_+ (\overline{\partial_+^{i+1} f_\alpha} \hat{Z}_\alpha) \qquad 3.29$$

$$= (-1)^{k-1} \delta^{i,k-1} \partial_+ |\hat{Z}|^2$$

Also

$$\begin{aligned} \overline{\partial_+^i f_\alpha} \partial_- \hat{Z}_\alpha &= \partial_- (\overline{\partial_+^i f_\alpha} \hat{Z}_\alpha) - (\overline{\partial_+^{i+1} f_\alpha} \hat{Z}_\alpha) = \\ &= (-1)^{k-1} \delta^{i,k-1} |\hat{Z}|^2 \end{aligned} \qquad 3.30$$

Hence, for $i = 0, \ldots k-1$,

$$\begin{aligned} \overline{\partial_+^i f_\alpha} D_+ D_- Z_\alpha &= \frac{1}{|\hat{Z}|} (-1)^{k-1} \delta^{i,k-1} \left[\partial_+ |\hat{Z}|^2 - \frac{1}{|\hat{Z}|^2} (\bar{\hat{Z}}_\alpha \partial_+ \hat{Z}_\alpha) \right] \\ &= \frac{1}{|\hat{Z}|} (-1)^{k-1} \delta^{i,k-1} (\overline{\partial_- \hat{Z}_\alpha} \hat{Z}_\alpha) = 0 \end{aligned} \qquad 3.31$$

Thus we see that both Z_α and $D_+ D_- Z_\alpha$ are in $\hat{H}_k$ and orthogonal to $\hat{H}_{k-1}$ - so they must be proportional.

$$D_+ D_- Z_\alpha = \lambda Z_\alpha \qquad 3.32$$

But

$$|D_-Z|^2 = \overline{D_-Z_\alpha} \; D_-Z_\alpha =$$
$$= \partial_+(\bar{Z}_\alpha \, D_-Z_\alpha) - \bar{Z}_\alpha \, D_+D_-Z_\alpha = -\lambda \qquad 3.33$$

Hence

$$D_+D_-Z_\alpha = -|D_-Z|^2 \; Z_\alpha \qquad 3.34$$

which shows that Z_α satisfies the equations of motion.

We see then how to obtain solutions of the equations of motion. We take any rational analytic vector $f_\alpha = f_\alpha(x_+) \in C^n$ (we could of course equally well start with $f_\alpha = f_\alpha(x_-)$). Then Z_α defined through 3.20 and 3.22 in terms of f_α and its derivatives satisfies the equations of motion.

It remains to be shown that Z_α has a finite action and to determine its value in terms of properties of f_α and its derivaties. This was done in ref. 7 - here we shall defer this discussion until the next section in which we shall first of all introduce a different but equivalent formulation of the problem of construction of solutions and only then discuss their properties.

4. Alternative formulation

Consider a vector $0 \neq g \in C^n$. Define an operator P_+ by

$$(P_+g)_\alpha = \partial_+g_\alpha - \frac{g_\alpha(\bar{g}_\beta\partial_+g_\beta)}{\bar{g}_\gamma g_\gamma} \qquad 4.1$$

and define its repeated action as

$$(P_+^k g)_\alpha = P_+(P_+^{k-1} g_\alpha) \qquad 4.2$$

It is then a matter of arithmetic to show that $\hat{Z}_\alpha$ of 3.20 is in fact given by

$$\hat{Z}_\alpha^{(k)} = (-1)^k (P_+^k f)_\alpha \qquad 4.3$$

Hence the P_+ operator can be thought of as a kind of a Bäcklund transformation operator (which here however does not introduce any further parameters).

A further insight can be gained if we introduce wedge products of f_α and its derivatives. We define

$$h^{(i)} = f \wedge \partial_+ f \wedge \partial_+^2 f \wedge \ldots \wedge \partial_+^i f \qquad i = 0, \ldots n-1 \qquad 4.4$$

Then as it is easy to check

$$P_+^k f \sim (h^{k-1})^+ h^k \qquad 4.5$$

where the dot product in 4.5 denotes the summation over all indices of h^{k-1}, and all but one of h^k. If f_α is analytic, so is $h^{(i)}$ - although it is an element of a larger-dimensional space. Thus 4.5 suggests a possible interpretation for the noninstanton ($k \neq 0$) solutions - they correspond to mixtures of instantons and antiinstantons. Both instantons and antiinstantons are elements of larger-dimensional spaces and are of special form (eqn. 4.4). Their special form allows their mixtures to be elements of C^n, corresponding to the solutions of the equations of motion of the CP^{n-1} model. Thus in a way 4.5 shows that all nonlinearities of the CP^{n-1} model are associated with the dimension of its manifold and when properly reinterpreted the equations of motion are just like in the instanton case, the Cauchy-Riemann relations for vectors $h^{(i)}$. We shall see later on that the specific form of the superposition of $h^{(i)}$ vectors as shown in 4.5 when generalized to non-adjacent $h^{(i)}$ leads to solutions of other Grassmannian models.

Let us mention some useful properties of $(P_+^k f)_\alpha$ (f_α - analytic) which follow directly from their definition or are very easy to prove

1) $P_+^n f_\alpha = 0$

2) $\overline{P_+^k f_\alpha}\, P_+^\ell f_\alpha = 0 \quad \text{if } k \neq \ell$

3) $\partial_-(P_+^k f_\alpha) = -P_+^{k-1} f_\alpha \dfrac{|P_+^k f|^2}{|P_+^{k-1} f|^2}$ $\qquad$ 4.6

4) $\partial_+\left(\dfrac{P_+^{k-1} f_\alpha}{|P_+^{k-1} f|^2}\right) = \dfrac{P_+^k f_\alpha}{|P_+^{k-1} f|^2}$

and taking $Z_\alpha = \dfrac{P_+^k f_\alpha}{|P_+^k f|}$

5) $D_+|P_+^k f| = 0$

6) $D_-|P_+^k f|^{-1} = 0$

In fact using these expression it is very easy to prove that $Z_\alpha = \dfrac{P_+^k f_\alpha}{|P_+^k f|}$ satisfies the equations of motion.

Next we proceed to calculate the expressions for the topological charge and for the action. Using the relations given above and the relation

$$D_\pm(af_\alpha) = (D_\pm a) f_\alpha + a(\partial_\pm f_\alpha) \qquad 4.7$$

we find that for $Z_\alpha = \frac{P_+^k f_\alpha}{|P_+^k f|}$

$$|D_-Z|^2 = \frac{|P_+^k f|^2}{|P_+^{k-1} f|^2}, \quad |D_+Z|^2 = \frac{|P_+^{k+1} f|^2}{|P_+^k \ f|^2} \qquad 4.8$$

On the other hand a few lines of algebra show that

$$\partial_+\partial_- \ell n|P_+^k f|^2 = \partial_+ \frac{1}{|P_+^k f|^2} \partial_-|P_+^k f|^2 = \partial_+ \frac{1}{|P_+^k f|^2} (\overline{\partial_+ P_+^k f} \cdot P_+^k f) =$$

$$= \frac{\overline{\partial_-\partial_+ P_+^k f_\alpha} \cdot P_+^k f_\alpha}{|P_+^k f|^2} + \frac{\overline{\partial_+ P_+^k f_\alpha} \cdot P_+^{k+1} f_\alpha}{|P_+^k \ f|^2} = |D_+Z|^2 - \qquad 4.9$$

$$\frac{\overline{\partial_+\left(P_+^{k-1} f_\alpha \frac{|P_+^k f|^2}{|P_+^{k-1} f|^2}\right)} \cdot P_+^k f_\alpha}{|P_+^k f|^2} = |D_+Z|^2 - |D_-Z|^2$$

Hence $Q = 2\partial_+\partial_- \ell n|P_+^k f|^2$. On the other hand as $S = Q + 2|D_-Z|^2$ and as

$$\left|D_-\left(\frac{P_+^{k+1} f}{|P_+^{k+1} f|}\right)_\alpha\right|^2 = \left|D_+\left(\frac{P_+^k f}{|P_+^k f|}\right)_\alpha\right|^2 \qquad 4.10$$

we see that

$$\frac{1}{2}S = \partial_+\partial_- \ln|P_+^k f|^2 + 2\sum_{i=0}^{k-1} \partial_-\partial_+ \ln|P_+^i f|^2 \qquad 4.11$$

Next we calculate the integrated values of the topological charge and of the action.
As

$$\partial_+\partial_- \ln|p|^2 = \frac{1}{4}\partial_\mu\partial_\mu \ln|p|^2 , \qquad 4.12$$

the use of the divergence theorem in two dimensions shows that

$$\int d^2x \ \partial_+\partial_- \ln|p|^2 = \pi\alpha \qquad 4.13$$

where $|p| \to |x|^\alpha$ as $|x| \to \infty$.

Here we have assumed that $|p|$ has no singularities except those at ∞.

It is easy to see that up to an overall constant

$$|P_+^k f|^2 \cong |h^k|^2/|h^{k-1}|^2 \qquad 4.14$$

and so

$$\partial_+ \partial_- \ln |P_+^k f|^2 = \partial_+\partial_- \ln|h^k|^2 - \partial_+\partial_- \ln|h^{k-1}|^2 . \qquad 4.15$$

Hence

$$Q^k(x) = 2(\partial_+\partial_- \ln|h^k|^2 - \partial_+\partial_- \ln|h^{k-1}|^2)$$

and

$$S^k(x) = 2(\partial_+\partial_- \ln |h^k|^2 + \partial_+\partial_- \ln|h^{k-1}|^2 .) \qquad 4.16$$

The vector f_α used in the construction of our solutions has analytical components which are rational functions. Due to the gauge invariance we can multiply all its components by an overall analytical factor - turning its components into analytic polynomials. Then the resultant h^k vectors have no singularities (except those at ∞) and so we can use the expression given in 4.13.

It is not too difficult to convince oneself that the condition of finite action is equivalent, up to a gauge transformation, to the choice of a polynomial form for f_α . It may happen that h^k being a polynomial vector exhibits an overall analytical polynomial factor. Then $\ln |h^k|^2$ exhibits a singularity which would have to be taken into account when calculating 4.13. However such an overall factor would introduce an overall factor in $P_+^k f$ and so will correspond to an overall phase in $Z_\alpha^k = P_+^k f_\alpha / |P_+^k f|^2$. On the other hand it would affect the behaviour of h^k at ∞ ; and it is not difficult to show that the contribution due to the singularity of $\ln|h^k|^2$ cancels the effect due to the increased behaviour at ∞ . Thus we shall disregard such factors and introduce

$$\alpha_i = \deg f \wedge \partial_+ f \wedge \ldots \wedge \partial_+^i f \quad \text{mod (overall factors)} \qquad 4.17$$

$$i = 0, \ldots n-1$$

Then from the discussion given above we see that

$$Q_k = 2\pi(\alpha_k - \alpha_{k-1}), \qquad \alpha_{k-1} = 0$$

$$4.18$$

$$S_k = 2\pi(\alpha_k + \alpha_{k-1}) ,$$

Observe that the vector h^{n-1} (if it has not vanished earlier) is uniquely determined and so all its x_+ dependence is in a form of an overall factor. Thus $\alpha_{n-1} = 0$ and so we see that for $k = n-1$ we have

$$Q_k =- S_k = -2\pi \alpha_{k-2} \qquad 4.19$$

and so $Z_\alpha^{n-1} = \dfrac{P_+^{n-1} f_\alpha}{|P_+^{n-1} f|}$ corresponds to antiinstantons.

Also it is clear that in the CP^1 case (n=2) there are only two levels - instantons and antiinstantons and the P_+ operator turns instantons into antiinstantons. Looking at the values of Q_k and S_k we see that the interpretation of the noninstanton solutions as corresponding to mixture states of instantons and antiinstantons is further reinforced by their values of the action and of the topological charge. We see that the mixture configurations are very special - not only do they satisfy equations of motion but their interaction action vanishes.

The noninstanton solutions have $Q_k \neq S_k$ and so for them the usual topological arguments guaranteeing their stability do not apply - hence they may be unstable and so not to correspond to local minima of the action. We shall show next that this is indeed the case. Let us take a solution z_α, for which $D_\pm Z_\alpha \neq 0$, and let us consider a small fluctuation φ_α around z_α. As a result of this fluctuation z_α field is modified to

$$Z'_\alpha = \sqrt{1-|\varphi|^2}\, Z_\alpha + \phi_\alpha \qquad 4.20$$

where

$$\bar{Z}_\alpha \varphi_\alpha = 0$$

The action for the field Z'_α is

$$S' = 2\int d^2x\,[\,|D'_+Z'|^2 + |D'_-Z'|^2\,] \qquad 4.21$$

where $D'_\pm$ is the usual covariant derivative written in terms of Z'_α. Since the topological charge is invariant under the introduction of this fluctuation,

$$S' = 2\int d^2x\, Q(x) + 4\int d^2x\, |D'_-Z'|^2 \qquad 4.22$$

Calculating to second order in the small fluctuation ϕ_α we have

$$|D'_-Z'|^2 = |D_-Z|^2 + |D_-\phi|^2 - |\phi|^2|D_-Z|^2 - |\bar{Z}_\alpha D_-\cdot\phi_\alpha + \bar{\phi}_\alpha\cdot D_-Z_\alpha|^2 \qquad 4.23$$

$$+ \overline{D_-Z_\alpha}\, D_-\phi_\alpha + \overline{D_-\phi_\alpha}\, D_-Z_\alpha$$

giving

$$S' = S + 4\int d^2x\, V(\phi) \qquad 4.24$$

where S is the action of z_α and

$$V(\phi) = |D_-\phi|^2 - |\phi|^2|D_-Z|^2 - |\bar{Z}_\alpha D_-\phi_\alpha + \bar{\phi}_\alpha D_-Z_\alpha|^2 + \overline{D_-\phi_\alpha}\cdot D_-Z_\alpha \qquad 4.25$$

$$+ D_-\phi_\alpha\cdot\overline{D_-Z_\alpha}$$

If we now choose

$$\phi_\alpha = \varepsilon\, D_+Z_\alpha \qquad 4.26$$

where ε is a small complex constant number, then as

$$D_- \phi_\alpha = \varepsilon D_- D_+ Z_\alpha = -\varepsilon |D_+ Z|^2 Z_\alpha \qquad 4.27$$

and

$$\bar{\phi}_\alpha D_- Z_\alpha = 0$$

we find that for this fluctuation

$$\begin{aligned} V(\phi) &= |\varepsilon|^2 |D_+ Z|^4 - |\varepsilon|^2 |D_+ Z|^2 |D_- Z|^2 - |\varepsilon|^2 |D_+ Z|^4 = \\ &= - |\varepsilon|^2 |D_+ Z|^2 |D_- Z|^2 < 0 \end{aligned} \qquad 4.28$$

showing that S' < S and so that the noninstanton solutions are unstable - i.e. they correspond to the saddle points of the action. Of course the fluctuation ϕ_α of 4.26 is not the only negative mode of the action. It is easy to see that in general ε does not always have to be a constant and we can take fluctuations in the direction of higher powers of the P_+ operator. Given the polynomial form of f_α such fluctuations do not take the fields outside their topological sector. However, the work on this point has not yet been completed. There is a case in which we have complete understanding of the fluctuations spectrum. This is the case of the CP^2 embedded O(3) instanton. That real solution of the CP^2 equations of motions was studied in ref. 12. In our present language it can be shown to result from the application of the P_+ operator to the following f

$$f_\alpha = (x_+^2 + 1,\ x_2^2 - 1,\ 2x_+). \qquad 4.29$$

In ref. (12) it was shown that the solution corresponding to this $P_+ f_\alpha$ is characterized by 6 negative modes and 16 zero modes. It is clear that all zero modes of a general $f_\alpha/|f|$ remain zero modes of $P_+^k f_\alpha/|P_+^k f|$ and it is very likely that $P_+^k f_\alpha/|P_+^k f|$ has no further modes. As the action of z^k is given by $(\alpha_k + \alpha_{k-1})2\pi$ and so corresponds to α_k instantons and α_{k-1} antiinstantons it was suggested in ref. (13) that the number of negative modes is probably given by

$$n(\alpha_k + \alpha_{k-1} - \alpha_o)$$

which would correspond to the situation in which the instability of the system of instantons and antiinstantons would effectively convert the equal number of zero modes of a system in which all antiinstantons were instantons ($S = Q = 2\pi(\alpha_k + \alpha_{k-1})$) into positive and negative modes. Such a picture would be true if the only role of the instanton-antiinstanton interaction were to introduce the instability but not to modify the action. All the cases studied support the correctness of this suggestion - the proof (or disproof) will have to wait until the spectrum of the fluctuation operator is better undestood.

Let us close this section with the observation that in addition to the P_+ operator we can also define the P_- operator

$$P_- g_\alpha = \partial_- g_\alpha - g_\alpha \frac{(\bar{g}_\beta \partial_- g_\beta)}{\bar{g}_\gamma \; g_\gamma} \qquad 4.30$$

and

that the role of this operator is exactly opposite to that of the P_+ operator.

As it is easy to check

$$P_- P_+ g_\alpha \sim g_\alpha \qquad 4.31$$

and so instead of starting with analytic vectors and the use of the P_+ operator we could have used antianalytic vectors and applied to them the P_- operator. Thus we see that the emerging picture is that of n levels of solutions (for CP^{n-1}) characterized by different values of the topological charge, but also by the same number of zero modes.

instantons 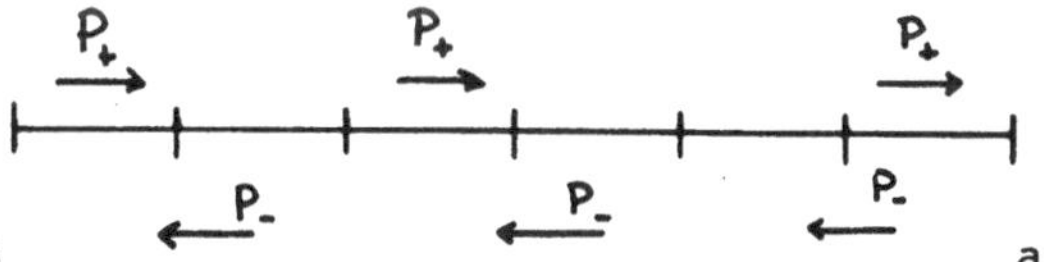 antiinstantons

The two extreme solutions correspond to instantons and antiinstantons respectively and the operators P_+ and P_- convert these solutions to each other. For a general enough f_α (i.e. when there are no special coincidences or degeneracies) the topological charge decreases by 2 with each application of the P_+ operator and the solution in the middle, (for n odd or the two in the middle for n even) have the largest value of the action and correspond to the largest number of instantons and antiinstantons. Clearly there exists a group theoretical explanation of the properties of this system of solutions.

5. Supersymmetric extensions

Of the several extensions of the CP^{n-1} model to include fermions the most interesting and most widely studied one is the supersymmetric extension of d'Adda et al. [14] . In that extension the z_α field is replaced by a superfield $\phi_\alpha(x,y,\theta_1,\theta_2)$ where θ_1, θ_2 are the two components of a real grassmannian spinor which defines the superspace extension of the twodimensional space-time.

The action of the supersymmetric extension of CP^{n-1} model is given by

$$S = \int d^2x d\theta_1 \, d\theta_2 [\overline{\partial'\phi_\alpha}\gamma_5 \partial'\phi_\alpha - \bar{\phi}_\alpha \partial' \phi_\alpha \gamma_5 \bar{\phi}_\beta \partial'\phi_\beta] \quad 5.1$$

where

$$\partial' = -i\partial_\theta + \not{\partial}\theta \quad 5.2$$

and where

$$\phi_\alpha = Z_\alpha + i\theta\chi_\alpha + \frac{i}{2}\theta\gamma_5\theta F_\alpha \quad 5.3$$

satisfies

$$\bar{\phi}_\alpha \phi_\alpha = 1 \, . \quad 5.4$$

In the conventions of d'Adda et al.[14]

$$\gamma_5 = \begin{pmatrix} 0 & 1 \\ -1 & 0 \end{pmatrix}, \qquad \theta = \begin{pmatrix} \theta_1 \\ \theta_2 \end{pmatrix}, \qquad \gamma_0 = \begin{pmatrix} 1 & 0 \\ 0 & -1 \end{pmatrix}, \quad \gamma_1 = \begin{pmatrix} 0 & 1 \\ 1 & 0 \end{pmatrix} \quad 5.5$$

and so $\theta\gamma_5\theta = 2\theta_1\theta_2$.

The spinors θ and χ anticommute, and

$$\theta\chi_\alpha = \theta_1\chi_{1\alpha} + \theta_2\chi_{2\alpha} \, .$$

The supersymmetric extension of the topological charge is given by

$$Q = i\int d^2x \, d\theta_1 d\theta_2 [\overline{\partial'\phi_\alpha} \, \partial'\phi_\alpha - (\bar{\phi}_\alpha \partial'\phi_\alpha)(\bar{\phi}_\beta \partial' \phi_\beta)] \quad 5.6$$

The Euler-Lagrange equations of the theory are

$$D'\gamma_5 D'\phi_\alpha + (\overline{D'\phi_\beta} \, \gamma_5 D'\phi_\beta)\phi_\alpha = 0 \quad 5.7$$

where

$$D' = \partial' - (\bar{\phi}_\alpha \partial'\phi_\alpha) \quad 5.8$$

and the supersymmetric generalisations of the instanton (antiinstanton) equations are

$$D'\phi_\alpha = \pm i \, \gamma_5 \, D'\phi_\alpha \quad 5.9$$

Introducing holomorphic (and antiholomorphic) $\theta_\pm$ [15] by

$$\theta_\pm = \theta_1 \pm i\theta_2 \quad 5.10$$

in addition to $x_\pm$ and then noticing that

$$\partial'_\pm = -i\partial_{\theta_\pm} + \theta_\pm \partial_\pm \quad 5.11$$

We find (following the discussion in section 2) that the equations of motion 5.6 can be rewritten as

$$D'_+D'_- \phi_\alpha + (\overline{D'_-\phi_\beta}\, D'_-\phi_\beta)\phi_\alpha = 0 \qquad 5.12a$$

or

$$D'_-D'_+\phi_\alpha + (\overline{D'_+\phi_\beta}\, D'_+\phi_\beta)\phi_\alpha = 0 \qquad 5.12b$$

where $D'_\pm$ is given by 5.7 in which ∂' is replaced by $\partial'_\pm$.

The supersymmetric extensions of the instanton equations are

$$D'_\mp \phi_\alpha = 0 \qquad 5.13$$

and so, like in the pure CP^{n-1} case, have a simple solution

$$\phi_\alpha = \frac{\omega_\alpha}{|\omega|} \qquad 5.14$$

where

$$\omega_\alpha = \omega_\alpha(x_+, \theta_+)$$

or 5.15

$$\omega_\alpha = \omega_\alpha(x_-, \theta_-)$$

respectively.

Working out expression 5.15 in terms of fields we find

$$\omega_\alpha = f_\alpha(x_+) + i\,\theta_+\, g_\alpha(x_+)$$

or 5.16

$$\omega_\alpha = f_\alpha(x_-) + i\,\theta_-\, g_\alpha(x_-)$$

The expressions 5.14 and 5.16 constitute supersymmetric extensions of the instanton (antiinstanton) solutions of the CP^{n-1} model. Expanding supersymmetric instanton solutions in terms of Z , $\chi_{\alpha i}$, F_α fields of 5.3 we find that

$$Z_\alpha = \frac{f_\alpha}{|f|}$$

$$\chi_{+\alpha} = g_\alpha - \frac{1}{2}\frac{(\bar{f}_\beta g_\beta)}{|f|^2} f_\alpha$$

$$\chi_{-\alpha} = -\frac{1}{2}\frac{(f_\beta \bar{g}_\beta)}{|f|^2} f_\alpha$$

and 5.17

$$F_\alpha = \frac{3}{2}\frac{(f_\beta \bar{g}_\beta)(\bar{f}_\gamma g_\gamma)}{|f|^4} f_\alpha - \frac{(\bar{g}_\beta g_\beta) f_\alpha}{|f|^2} - \frac{(f_\beta \bar{g}_\beta) g_\alpha}{|f|^2}$$

If following d'Adda et al. we transform $\chi_{\alpha\pm}$ to the $\psi_\pm$ fields defined by

$$\psi_{\alpha\pm} = \chi_{\alpha\pm} - (\bar{Z}_\beta \chi_{\beta\pm}) Z_\alpha \qquad 5.18$$

we find that the supersymmetric instanston solutions correspond to

$$\psi_{\alpha+} = g_\alpha - \frac{(\bar{f}_\beta g_\beta) f_\alpha}{|f|^2} \qquad 5.19$$

$$\psi_{\alpha-} = 0$$

Here and above we resolved the spinors χ_α and ψ_α in terms of eigenstates of γ_5.

$$\psi_\alpha = \begin{pmatrix} \psi_{\alpha 1} \\ \psi_{\alpha 2} \end{pmatrix} = \begin{pmatrix} 1 \\ i \end{pmatrix} \psi_{\alpha+} + \begin{pmatrix} 1 \\ -i \end{pmatrix} \psi_{\alpha-} \qquad 5.20$$

Having discussed the solutions to the first order equations 5.13 we now turn our attention to the solutions of 5.11 or 5.12 . It is not difficult to show that almost all the steps of the previous sections go through if we take $\omega_\alpha(x_+,\theta_+)$ instead of f_α and in the place of ∂_+ consider ∂'_+. However as ∂'_+ interchanges bosonic and fermionic degrees of freedom and in our construction of the basis in the space $\hat{H}_k$ (recall $f, \partial_+ f, \partial_+^2 f, \ldots \partial_+^k f$), we need bosonic quantities, we have to introduce arbitrary constant nonzero spinors $\varepsilon_0, \varepsilon_1 \ldots \varepsilon_{\frac{n-1}{2}}$ and then take for the basis

$$\omega, \ \varepsilon_0 \partial'_+ \omega, \quad \partial'^2_+ \omega \ , \quad \varepsilon_1 \partial'^3_+ \omega \ , \ \partial'^4_+ \omega, \ \ldots$$

To proceed further we calculate various wedge products and introduce

$$h^k = \omega \wedge \varepsilon_0 \partial'_+ \omega \wedge \partial'^2_+ \omega \wedge \ldots \varepsilon_{\frac{k-1}{2}} \partial'^k_+ \omega \qquad 5.21$$

for k odd

and

$$h^k = \omega \wedge \varepsilon_0 \partial'_+ \omega \wedge \partial'^2_+ \omega \wedge \ldots \wedge \partial'^k_+ \omega \qquad 5.22$$

for k even

and then define

$$\hat{\phi}_i = (h^{k-1})^+ h^k / h^{k-1+} h^{k-1} \qquad 5.23$$

We can then take

$$H_i = \{ \hat{\phi}_0, \hat{\phi}_1 \ldots \hat{\phi}_i \}$$

and verify that

$$\varepsilon \partial'_+ \phi^i \in H_{i+1}$$

and

$$\varepsilon\partial_-'\phi^i \in H_{i-1}$$

Using these properties and the fact that the set $\{\phi_i\}$ is an orthogonal system we find

$$\partial_-'\hat{\phi}^{2k+1}\bar{\varepsilon}_k = -\frac{|\hat{\phi}^{2k-1}|^2}{|\hat{\phi}^{2k}|^2}\hat{\phi}^{2k}$$

$$\partial_+'\left(\frac{\hat{\phi}^{2k+1}}{|\hat{\phi}^{2k+1}|^2}\right) = \varepsilon_k \frac{\hat{\phi}^{2k+2}}{|\hat{\phi}^{2k+1}|^2} \qquad 5.24$$

$$\partial_-'\hat{\phi}^{2k} = -\bar{\varepsilon}_{k-1}\frac{|\hat{\phi}^{2k}|^2}{|\hat{\phi}^{2k-1}|^2}\hat{\phi}^{2k-1}$$

and

$$\varepsilon_k\partial_+ \frac{\hat{\phi}^{2k}}{|\hat{\phi}^{2k}|^2} = \frac{\hat{\phi}^{2k+1}}{|\hat{\phi}^{2k}|^2}$$

which all together are the analogues of the relations 4.6. Also taking

$$\phi^k = \frac{\hat{\phi}^k}{|\hat{\phi}^k|} \qquad 5.25$$

and

$$D_\pm' = \partial_\pm' - \bar{\phi}^k\partial_\pm'\phi^k \qquad 5.26$$

we find

$$D_+' \, |\hat{\phi}^k| = 0 \text{ and } \quad D_-' \, |\hat{\phi}^k|^{-1} = 0 \qquad 5.27$$

which are the analogoues of 4.7. Then it is a matter of algebra to show that ϕ^k of 5.25 satisfies the equation 5.11 and as two fields which differ from each other by an overall grassmannian valued number

$$a+\theta_+b + \theta_-c + \theta_+\theta_-d$$

are considered to be equivalent it is easy to see that there are no new solutions in the supersymmetric CP^1 case and, like in the CP^{n-1} case, the theory has further "new" solutions for $n > 2$. Their interpretation is again the same - they correspond to the mixture states of supersymmetric instantons and antiinstantons.

To look at the field content we write $\hat{\phi}$ of 5.25 as

$$\hat{\phi}_k = a_k + \theta_+b_k + \theta_-c_k + \theta_+\theta_-d_k \qquad 5.28$$

and then a_k corresponds to a bosonic contribution.

We can then calculate the values of the action and the topological charge - it is easy to see that only the bosonic terms contribute. Also we can consider the fields $\psi_{\alpha\pm}$ of d'Adda et al. and see what expressions our solutions 5.28 imply for them. However as the procedure of taking wedge products mixes f_α and g_α (the bosonic and fermionic degrees of freedom of the supersymmetric instantons) the interpretation of the obtained solutions is a bit difficult - this problem is still under consideration.

Instead of continuing this line of discussion we now turn our attention to the simpler problem of a fermion in the fixed background field of a bosonic CP^{n-1} solution[16]. The equation to solve is

$$\not{D}\psi_\alpha - (\bar{z}_\beta \not{D}\, \psi_\beta) z_\alpha = 0 \tag{5.29}$$

together with the constraint

$$\bar{z}_\alpha\, \psi_\alpha = 0 \tag{5.30}$$

It is easy to see that this problem comes from the original supersymmetric Lagrangian (5.1) written out, after integration over θ_1 and θ_2, in terms of z_α and ψ_α fields of d'Adda et al.[14]. If one then drops all nonlinear terms in ψ_α and $\bar{\psi}_\alpha$ in the equations of motion one obtains the equations of motion of the CP^{n-1} model - for the z_α field and the equations 5.29 for the ψ_α field. If like in 5.20 we resolve ψ_α in terms of eigenstates of γ_5 we find that 5.29 is equivalent to

$$D_\pm \psi_\alpha^\pm = \lambda_\pm z_\alpha \tag{5.31a}$$

with the constraint

$$\bar{z}_\alpha\, \psi_\alpha^\pm = 0 \tag{5.31b}$$

where $\lambda_\pm$ are some functions of x_+ and x_-.

If we denote by ξ_+ and ξ_- the numbers of independent $\psi^\pm$ solutions which are normalisable on the sphere i.e.

$$\int d^2x \frac{1}{1+x^2} |\psi^\pm|^2 < \infty \tag{5.32}$$

then the Atiyah-Singer index theorem tells us that

$$\xi_+ - \xi_- = -nQ \tag{5.33}$$

where Q is the topological charge of the background z_α field. If we return back to the instanton case then it can be shown that 5.19 is also the only normalisable solution of 5.31a - hence ξ_+ is zero for an instanton background and the index theorem (5.33) is satisfied in a minimal way.

We now take a general solution for the background field - i.e. corresponding to $\hat{z}_\alpha^{(k)} \sim P_+^k f_\alpha$ - and consider first the positive helicity equation. A ψ_α^+ field, orthogonal to $\hat{z}_\alpha^{(k)}$ can be written as

$$\psi_\alpha^+ = |\hat{z}^k| \sum_{i\neq k} \frac{\overline{(\hat{z}^i_\beta\, g^+_\beta)}}{|\hat{z}^i|^2}\, \hat{z}^{(i)}_\alpha \qquad 5.34$$

for some vector g^+ since $\hat{z}^k$ $(k=0,\ldots n-1)$ form an orthogonal system. Since

$$\partial_+ \sum_{i=0}^{k-1} \frac{\overline{\hat{z}^i_\beta}\, \hat{z}^i_\alpha}{|\hat{z}^i|^2} = \frac{\overline{\hat{z}^{k-1}_\beta}\, \hat{z}^k_\alpha}{|\hat{z}^{k-1}|^2} \qquad 5.35$$

as can be easily checked using relations 4.6, we see that ψ_α^+ satisfies 5.31 if $g^+_\alpha = g^+_\alpha(x_-)$ only and if $\overline{\hat{z}^i_\beta}\, g^+_\beta = 0$ for $i > k-1$ (this last condition comes from the conditions of normalisability of ψ_α^+ - they cannot be satisfied for any form of g^+_β). Thus the general form of ψ_α^+ satisfying 5.31 is given by

$$\psi_\alpha^+ = |\hat{z}^k_\alpha| \sum_{i=0}^{k-1} \frac{\overline{(\hat{z}^i_\beta g^+_\beta)}}{|\hat{z}^i|^2}\, \hat{z}^i_\alpha \qquad g^+_\beta = g^+{}_\beta(x_-) \qquad 5.36$$

Using similar arguments one finds that a general solution for ψ_α^- is given by

$$\psi_\alpha^- = \frac{1}{|\hat{z}^k|} \sum_{i=k+1}^{n-1} \frac{\overline{(\hat{z}^i_\beta\, g^-_\beta)}}{|\hat{z}^i|^2}\, \hat{z}^i_\alpha \qquad g^-_\beta = g^-_\beta(x_+)\ . \qquad 5.37$$

Of course the normalisation conditions are not automatically satisfied - it remains to determine the most general forms of g^+_β and g^-_β which give ψ_α^+ and ψ_α^- satisfying the normalisation condition 5.32 and then counting the number of independent parameters in ψ_α^+ and ψ_α^- to determine ξ_+ and ξ_-.

For the norms of $\psi_\alpha^\pm$ we have

$$|\psi^+|^2 = |\hat{z}^k|^2 \sum_{i=0}^{k-1} \frac{|\overline{\hat{z}^i_\beta g^+_\beta}|^2}{|\hat{z}^i|^2} \qquad 5.38$$

$$|\psi^-|^2 = \frac{1}{|z^k|^2} \sum_{i=k+1}^{n-1} \frac{|\overline{\hat{z}^i_\beta}\, g^-_\beta|^2}{|\hat{z}^i|^2} \qquad 5.39$$

If we first consider the case of ψ_α^+, the functions $\overline{\hat{z}^i_\beta}\, g^+_\beta$ $i=0,\ldots k-1$ are independent. It can be shown that the maximal degree in $|x|$ as $|x| \to \infty$ is deg $\hat{z}^i\, g^+ = Q_i - Q_k - 1$ and so the number of independent parameters is $Q_i - Q_k$. (The discussion is complicated - first we show that g^+ is such that $\overline{\hat{z}^i_\beta}\, g^+_\beta$ is a polynomial. Then, for $i = 0$ $\overline{\hat{z}^i_\beta}\, g^+_\beta$ is a polynomial in x_- and so contains $Q_0 - Q_k$ independent parameters; for $i \neq 0$ we find that as $\partial_+ \overline{\hat{z}^i_\beta}\, g^+_\beta = -\,(|\hat{z}^i|^2/(|\hat{z}^{i-1}|^2)\ \overline{\hat{z}^{i-1}_\beta} g^+_\beta$, which are already counted and so only ∂_- derivatives give independent quantities - thus showing that there are $Q_i - Q_k$ independent parameters).

Hence

$$\xi_+ = \sum_{i=0}^{k-1} (Q_i - Q_k) = \alpha_{k-1} - kQ_k \qquad 5.40$$

Considering the case of ψ_α^- we find that $\deg \frac{\overline{\hat{z}_\alpha^i}\, g_\alpha^-}{|\hat{z}^i|^2} = Q_k - Q_i - 1$ and following a similar reasoning as for ψ_α^+ we find that there are exactly $Q_k - Q_i$ independent parameters. Hence

$$\xi_- = \sum_{i=k+1}^{n-1} (Q_k - Q_i) = \alpha_k + (n-k-1)Q_k \qquad 5.41$$

and we see that the index theorem is indeed satisfied:

$$\xi_+ - \xi_- = \alpha_{k-1} - kQ_k - \alpha_k - (n-k-1)Q_k = -Q_k - (n-1)Q_k = -nQ_k \qquad 5.42$$

as required.

Let us finish this section by pointing out that had we considered the problem of solving the Dirac equation corresponding to the case when the fermion field is coupled to the Z_α field of the CP^{n-1} model in a minimal way we would have

$$\not{D}\psi_\alpha = 0 \qquad 5.43$$

without any constraint on ψ_α. Then the general solutions of the corresponding $\psi_\alpha^\pm$ fields would be

$$\psi_\alpha^+ = |\hat{z}^k| g_\alpha^+(x_-) \qquad 5.44$$

and

$$\psi_\alpha^- = \frac{1}{|\hat{z}^k|} g_\alpha^-(x_+) \qquad 5.45$$

Then if $Q_k > 0$ we find $\xi_+ = 0$ and $\xi_- = nQ_k$ and for $Q_k < 0$ we find $\xi_- = 0$, $\xi_+ = -nQ_k$ i.e. the index theorem is fulfilled in a minimal way.

Finally, in the case of a CP^{n-1} solutions Z_α which is an embedding of a lower dimensional model we have an intermediate situation. For the projection of a fermion field on the subspace spanned by the solution the discussion proceeds as in the case of the supersymmetric Dirac equation, for the projection on the orthogonal subspace the discussion of the simple Dirac equation applies.

6. Grassmannian models

Grassmannian models[17] are simple generalisations of CP^{n-1} models. Instead of 2.1 we take

$$\mathcal{L} = (\overline{D_\mu z^i_\alpha} D_\mu z^i_\alpha) \quad \alpha = 1...n, \ i=1...p. \qquad 6.1$$

where

$$D_\mu z^i_\alpha = \partial_\mu z^i_\alpha - z^j_\alpha (\overline{z^j_\beta} \partial_\mu z^i_\beta) \qquad 6.2$$

and the constraint on z^i_α fields is

$$\overline{z^i_\alpha} z^j_\alpha = \delta_{ij}, \quad 1 \leq i,j \leq p \qquad 6.3$$

Thus the model exhibits a U(n) global symmetry, U(p) local symmetry and two fields z^i_α and $z^{i'}_\alpha$ which are related to each other by a U(p) local transformation are said to be gauge related and so are considered to be equivalent:

$$z'^i_\alpha \simeq z^i_\alpha \quad \text{if } z'^i_\alpha = U_{ij}(x,y) z^j_\alpha \qquad 6.4$$

where $U_{ij} \in U(p)$.

Clearly the CP^{n-1} model is a special case of these models - it corresponds to p=1. The equations of motion of these models which are a generalisation of 2.5 can be written as

$$D_\mu D_\mu z^i_\alpha + \overline{D_\mu z^j_\beta} D_\mu z^j_\beta z^i_\alpha = 0$$

or explicitly 6.5

$$\partial_\mu \partial_\mu z^i_\alpha - 2\partial_\mu z^k_\alpha \overline{z}^k_\gamma \partial_\mu z^i_\gamma = \lambda_{ij} z^j_\alpha$$

together with the constraint

$$\overline{z^i}_\alpha z^j{}_\alpha = \delta_{ij} .$$

The first order equations (for instantons) are

$$D_\pm z^i_\alpha = 0 \qquad 6.6$$

and it is easy to find their solutions. To find them one proceeds as follows. First take a set of p analytical vectors f^i_α i=1...p. Then form

$$h^k = f^1 \wedge f^2 \wedge \ ... \wedge f^k \qquad 6.7$$

and consider a set of $\hat{z}^i_\alpha$ obtained from

$$\hat{z}^{(i)}_\alpha = (h^{i-1})^+ h^i \qquad 6.8$$

and

$$\hat{z}_\alpha^{(1)} = h^{(1)} \qquad 6.9$$

Then it is easy to show that

$$z_\alpha^{(i)} = \frac{\hat{z}_\alpha^{(i)}}{|\hat{z}^{(i)}|} \qquad 6.10$$

satisfy 6.6 i.e. are instanton solutions of the grassmannian models. In fact it can be shown that this construction gives all instanton solutions (up to a gauge transformation).

What about solutions of the general equations of motion - i.e. 6.5? We have found some solutions of these equations and we will discuss them below. However, so far we have been unable to find all solutions of these equations or to convince ourselves that we have found them all. So let me show what solutions we have and discuss their properties.

For simplicity we shall discuss only the case of p=2. It is quite easy to generalise our discussion to the case p > 2.

We have essentially 3 classes (types) of solutions

<u>The first class</u>:

<u>Observe</u> that as

$$P_+^i f_\alpha \; P_+^j f_\alpha = 0 \quad \text{for } i \neq j$$

the choice of

$$z^i{}_\alpha = \frac{P_+^{j_i} f_\alpha}{|P_+^{j_i} f|} \qquad \begin{array}{l} 0 \leq j_i \leq n-1 \\ j_i \neq j_k \quad \text{for } i \neq k \end{array} \qquad 6.11$$

automatically satisfies the conditions of orthogonality of $z^i{}_\alpha$. Then it is a matter of algebra to check that $z^i{}_\alpha$ also satisfies the equation of motion 6.5. It is easy to notice that the choice of $j_1=0$, $j_2=1$ corresponds to an instanton solution - corresponding to $f^1=f$, $f^2 = \partial_+ f$, and that other choices do not correspond to instantons (the choice $f^1 = P_+^{n-1} f$, $f^2 = P_+^{n-2} f$ corresponds to antiinstantons). One can calculate the action and the topological charge of these solutions. We find

$$Q = \sum_i Q_{j_i} \qquad 6.12$$

for the topological charge and either

$$S = Q_{j_i} + S_{j_{i-1}} \qquad \text{if } j_{i+1} = j_{i-1} \qquad 6.13$$

or

$$S = S_{j_i} + S_{j_{i+1}} \qquad \text{if } j_{i+1} \neq j_{i-1} \qquad 6.14$$

These rules generalise in a straightforward way when p > 2 .Let us remark at this point that as the grassmann-valued fields are elements of $\frac{U(n)}{U(p)\ U(n-p)}$ an equivalent description of the model based on the Lagrangian 6.1 is provided by the complement of Z^i_α - i.e. again by the Lagrangian 6.1 in which $i = p+1, \ldots n$ together with the corresponding constraint (6.3). In the case of our solutions (6.11) such a description is given by the choice of $Z_{i\alpha}$ based on the remaining elements from the set $(f, P_+f, \ldots P_+^{n-1}f)$. It is easy to see that this complementary description has the same value of the action and an opposite value of the topological charge. It is also easy to see that the solutions of the grassmannian models discussed above are related to taking scalar products of nonadjacent h_i in 4.5 as mentioned in section 4.

Second class

A second class of solutions comes from the observation that we can take for Z^i_α some combinations of terms of previous solutions. Namely we can take

$$Z^1_\alpha = \frac{P^i_+ f_\alpha}{|P^i_+ f|} \quad \text{for some (large) } i \qquad 6.15$$

and also

$$Z^2_\alpha = \frac{\hat{Z}^2_\alpha}{|\hat{Z}^2|}\,, \qquad \hat{Z}^2_\alpha = \sum_{k=0}^{i-2} a_k\, P^k_+ f_\alpha \qquad 6.16$$

The constraint equations are now satisfied, but the equations of motion are not.

However as $P^k_+ f_\alpha$ is a linear combination of $f_\alpha, .., \partial^k_+ f_\alpha$ we can rewrite $\hat{Z}^2_\alpha$ as

$$\hat{Z}^2_\alpha = \sum_{k=0}^{i-2} b_k \partial^k_+ f_\alpha \qquad 6.17$$

for some b_k.

On the other hand it is easy to check that Z^1_α and Z^2_α satisfy the equations of motion if all b_k are analytic - i.e. $\partial_- b_k = 0$. (This in turn can be reexpressed as a condition on a_k). These solutions are more general than those of the first class as they involve, in addition to f_α, also arbitrary analytic functions b_k. Further solutions can be obtained from 6.15 and 6.16 by applying the P_+ operator to both or to either of them - the new vectors Z^i_α will constitute a solution if the power of P_+ in 6.15 is at least two units larger than the largest power of P_+ in 6.16.

A third class of solutions can be described most easily in terms of wedge products of vectors - like those in 6.7, and is a simple generalisation of the procedure described in section 4. We start with $H^1 = f \wedge g$ which as we have shown before corresponds to instantons. We define

$$H^1 = f \wedge g \,,$$
$$H^2 = f \wedge g \wedge \partial_+ f \wedge \partial_+ g, \qquad 6.18$$
$$H^3 = f \wedge g \wedge \partial_+ f \wedge \partial_+ g \wedge \partial_+^2 f \wedge \partial_+^2 g,$$
etc.

and then we consider a series of twoplanes

$$H^1, \; H^{1+}H^2, \; H^{2+}H^3 \;, \; H^{i+} \; H^{i+1}, \; \ldots \qquad 6.19$$

which are a natural generalisation of the $P_+^k f$ of 4.5. Then it is a matter of simple although tedious algebra to show that the twoplanes of 6.19, when appropriately turned into $Z_{k\alpha}$, satisfy the equations of motion 6.5.

So far we have not been able to understand completely the mutual overlap of all these classes of solutions - nor to determine whether all solutions can be shown to belong of one of these classes. Clearly for the last class the results are different depending on whether n is even or odd (in the first case the last element of the set in 6.19 corresponds to antiinstantons, while for n odd we seem to in addition to the construction given before have to repeat the construction using also antianalytic vectors). This problem is currently under active consideration and we hope to have a better understanding of it in the near future.

7. Open questions

Clearly a lot of work remains to be done. So far as the classical solutions are concerned we still have to understand more completely both the supersymmetric and the grassmannian models.

Then we should try to understand the quantum corrections to these classical results. Of course the existence of negative modes complicates the discussion - but presumably we should be able to find an appropriate analytical continuation. In the instanton case Berg and Lüscher and Fateev, Frolov and Schwarz [18] showed that the quantum corrections to the classical solutions can be described in terms of a gas of instanton quarks. It will be interesting to see when the corrections to the noninstanton solutions are determined what impact these solutions have on the properties of this gas. Perhaps as a result of these corrections the properties of the gas in all CP^{n-1} models become more alike thus giving more credibility to the 1/n expansion.

Also there are several other important questions that the new solutions raise of which the most important is perhaps the question whether a similar construction can be found for the nonabelian gauge theories and the HP^{n-1} models. Unfortunately at the moment, this question, like many others, remains unanswered.

Acknowledgements

I would like to thank Dr. A.M. Din for his collaboration with me on all the topics discussed in these lectures. Also I would like to thank E.F. Corrigan, D.B. Fairlie, M. Günaydin, J. Lukierski, I. Singer, R. Stora and many others for discussions and constructive criticisms and J. Hietarinta and C. Montonen for their invitation to give these lectures at the Symposium in Tvärminne and their hospitality in Finland.

References

1. M.F. Atiyah, V.G. Drinfeld, N.J. Hitchin and Yu.I. Manin - Phys. Lett. 65A, 185 (1978)
E. Corrigan and P. Goddard - Springer Lecture Notes in Physics 129, 14, Springer Verlag (1980)
2. E.B. Bogomolny - Sov. J. Nucl. Phys. 24, 449 (1976)
3. H. Osborn - Nucl. Phys. B159, 281 (1979)
B. Berg and M. Lüscher - Nucl. Phys. B160, 281 (1979)
E. Corrigan, P. Goddard, H. Osborn and S. Templeton - Nucl. Phys. B159, 469 (1979)
4. H. Eichenherr - Nucl. Phys. B146, 215 (1978)
V. Golo and A. Perelomov - Phys. Lett. 79B, 112 (1978)
5. A. d'Adda, P. di Vecchia and M. Lüscher - Nucl. Phys. B146, 63 (1978)
6. A.M. Din and W.J. Zakrzewski - Nucl. Phys. B174, 397 (1980)
7. A.M. Din and W.J. Zakrzewski - Phys. Lett. 95B, 419 (1980)
8. A.M. Din, J. Lukierski and W.J. Zakrzewski - in preparation
9. A.M. Din and W.J. Zakrzewski - in preparation
10. M.J. Borchers and W.D. Garber - Comm. Math. Phys. 72, 77 (1980)
11. J. Barbosa Trans. Am. Math. Soc. 210, 75 (1975)
12. A.M. Din and W.J. Zakrzewski - Lett. Nuovo Cimento 28, 121 (1980)
13. A.M. Din and W.J. Zakrzewski - Lapp preprint TH-26-(1980) - to appear in Nuclear Phys. B.

14. A. d'Adda, P. di Vecchia and M. Lüscher - Nucl. Phys. B152, 125 (1978)

15. P. di Vecchia and S. Ferrara - Nucl. Phys. B130, 93 (1977)
J. Lukierski - private communication

16. A.M. Din and W.J. Zakrzewski - Lapp preprint TH-28 (1980)

17. E. Brezin, C. Itzykson, J. Zinn-Justin and J.B. Zuber - Phys. Lett. 82B, 442 (1979)
S. Duane - Nucl. Phys. B168, 32 (1980)

18. B. Berg and M. Lüscher - Comm. Math. Phys. 69, 57 (1979)
V.A. Fateev, I.V. Frolov and A.S. Schwarz - Nucl. Phys. B154, 1 (1979).

GEOMETRICAL ANALYSIS OF INTEGRABLE SIGMA MODELS

H. Eichenherr

CERN - Geneva

Contents

INTRODUCTION

In recent years, efforts to solve models of quantum field theory exactly have had some beautiful successes. The most systematic, and most ambitious, way to deal with exactly integrable quantum field theories is the quantum inverse scattering method. In a rather direct way, this formalism uses the fact that the corresponding classical models are completely integrable. This property in turn is expressed most systematically in the classical inverse scattering method.

The interplay of classical and quantum integrability motivated most of the investigations of integrable classical systems carried out in recent years. By now, there exists a large variety of two dimensional models sharing properties such as the existence of solitons, of higher conservation laws, or, presumably most important, of a linear Lax representation. In these lectures I shall confine my attention to a particular class of models: the generalized non-linear σ models (NLSM). This class, containing several models of special interest like the S^N, $\mathbb{C}P^N$ and principal (chiral) models and showing all types of inner symmetry groups, admits a unified, "streamlined" treatment of integrability questions. Moreover, it is related to various other classes of models like the generalized sine-Gordon systems, the generalized Toda lattices and the Gross-Neveu models. The NLSM have been investigated using notions and methods of differential geometry[1)-4)]. In fact, the structure of the NLSM mainly depends on the geometrical properties of groups entering, for example, as isometry or stabilizer groups of certain manifolds. Thus the NLSM are particularly suitable for looking systematically at the interplay of classical dynamics, geometric structures and integrability properties.

These notes review this development, starting in Chapter 1 with a brief summary of well-known properties of the simplest example, the S^2 σ model. In Chapter 2, a sufficiently general formulation of the NLSM on a Riemannian homogeneous space is given. In Chapter 3, I explain the central statement that all NLSM on symmetric spaces possess a linear Lax representation (yielding higher non-local conservation laws). After some mathematical considerations concerning symmetric spaces, the general definition of the reduced system (the "generalized sine-Gordon equation") corresponding to the NLSM on a symmetric space is formulated in Chapter 5, while Chapter 6 finally deals with the subtle question of the existence of higher local conservation laws.

I have tried to restrict myself to the main arguments, leaving out most of the mathematical and technical details; the general theory is illustrated throughout by rederiving the respective statements for the simplest non-trivial example, the S^2 model. For both mathematical rigor and more complicated (sometimes also more significant) examples, the reader is referred to the original articles[1)-4)] and the reviews[5),6)].

1. INTEGRABILITY PROPERTIES OF THE CLASSICAL S^2 NON-LINEAR σ MODEL

Our prototype and main example[7)] is defined in terms of a real field vector $\vec{q}(x^0,x^1)$ taking values on the two dimensional sphere S^2:

$$\vec{q}^{\,2}(x^0,x^1) = \sum_{i=1}^{3} |q^i(x^0,x^1)|^2 = 1 \tag{1.1}$$

Incorporating this constraint by introducing a Lagrange multiplier, the dynamics is given by the action

$$S = \frac{1}{2}\int d^2x \left\{ \partial_\mu \vec{q}\, \partial^\mu \vec{q} + \lambda(\vec{q}^{\,2}-1) \right\} \tag{1.2}$$

and the field equations

$$\Box \vec{q} + (\partial_\mu \vec{q}\, \partial^\mu \vec{q})\vec{q} = 0 \quad , \quad \vec{q}^{\,2} = 1 \quad . \tag{1.3}$$

Action and field equations show that this classical field theory is invariant under Lorentz and conformal transformations and under global $O(3)$ rotations of $\vec{q}$. The rotation invariance implies the conservation of the Noether current ($\vec{q}$ denoting a column vector)

$$\partial_\mu j^\mu = 0 \quad , \quad j_\mu = \vec{q}(\partial_\mu \vec{q})^T - (\partial_\mu \vec{q})\vec{q}^{\,T} \quad . \tag{1.4}$$

The following system of differential equations[7)] plays a central role in the classical S^2 model:

$$\begin{aligned} \partial_\xi U^{(\gamma)} &= (1-\gamma^{-1})\, U^{(\gamma)} j_\xi \\ \partial_\eta U^{(\gamma)} &= (1-\gamma)\, U^{(\gamma)} j_\eta \end{aligned} \qquad \gamma \in \mathbb{R} \quad . \tag{1.5}$$

Here, ξ and η denote light cone co-ordinates:

$$\xi = \frac{1}{2}(x^0 + x^1) \quad , \quad \eta = \frac{1}{2}(x^0 - x^1)$$

$$j_\xi = j_0 + j_1 \quad , \quad j_\eta = j_0 - j_1 \quad .$$

(1.5) is a system of differential equations for an O(3) valued function $U^{(\gamma)}(\xi,\eta)$. Given the components j_ξ and j_η of the conserved Noether current [taking values in the Lie algebra o(3) of O(3)], (1.5) determines the $U^{(\gamma)}$ up to normalization provided both equations are compatible. But the compatibility condition, obtained by cross differentiation, is exactly the equation of motion, hence we may consider (1.5) as a Lax representation of the S^2 σ model.

Unfortunately, it has not yet been possible to solve (1.3) via (1.5) in the sense of the inverse scattering method. For reasonable boundary conditions, i.e., $j_\mu \to 0$ as $|x^1| \to \infty$, the $U^{(\gamma)}$ in (1.5) are asymptotically constant and not plane waves. So it is unclear in which way an inverse scattering formalism as a generalized Fourier transformation[8] can be constructed around (1.5).

Nevertheless, (1.5) can be used to derive an infinite set of higher continuity equations[9] starting with

$$\partial_\eta j_\xi + \partial_\xi j_\eta = 0 \tag{1.6.1}$$

$$\partial_\eta \left\{ \tfrac{1}{2}\Big[\int_{-\infty}^{\xi} d\xi' j_\xi(\xi',\eta), j_\xi(\xi,\eta)\Big] - j_\xi(\xi,\eta)\right\} + \partial_\xi \left\{ \tfrac{1}{2}\Big[\int_{-\infty}^{\xi} d\xi' j_\xi(\xi',\eta), j_\eta(\xi,\eta)\Big]\right\} = 0 . \tag{1.6.2}$$

(1.6.1) is the (local) isospin conservation law (1.4), while (1.6.2) and all the following ones are non-local conservation laws involving more and more integrations.

The S^2 σ model is closely related to the sine-Gordon theory[7]. Defining

$$\cos\alpha = \frac{(\partial_\xi \vec{q}\, \partial_\eta \vec{q})}{\|\partial_\xi \vec{q}\| \|\partial_\eta \vec{q}\|} , \tag{1.7}$$

(1.3) implies the equations

$$\begin{aligned} \Box\alpha + \|\partial_\xi \vec{q}\| \|\partial_\eta \vec{q}\| \sin\alpha &= 0 \\ \partial_\eta \|\partial_\xi \vec{q}\| &= 0 \\ \partial_\xi \|\partial_\eta \vec{q}\| &= 0 . \end{aligned} \tag{1.8}$$

For reasons to become clear later (cf. Chapter 5), (1.8) has been called the "reduced S^2 system". Equations (1.8) are the compatibility conditions of the Lax pair

$$\partial_\xi g^{(\gamma)} = g^{(\gamma)} \begin{pmatrix} 0 & -\gamma^{-1}\|\partial_\xi \vec{q}\| & 0 \\ \gamma^{-1}\|\partial_\xi \vec{q}\| & 0 & \partial_\xi \alpha \\ 0 & -\partial_\xi \alpha & 0 \end{pmatrix} \tag{1.9}$$

$$\partial_\eta g^{(\gamma)} = \gamma g^{(\gamma)} \begin{pmatrix} 0 & -\|\partial_\eta \vec{q}\| \cos\alpha & -\|\partial_\eta \vec{q}\| \sin\alpha \\ \|\partial_\eta \vec{q}\| \cos\alpha & 0 & 0 \\ \|\partial_\eta \vec{q}\| \sin\alpha & 0 & 0 \end{pmatrix}$$

Again, $g^{(\gamma)}$ is a one parameter family of O(3) valued functions, which, by the principle of "gauge equivalence"[10], are closely related to the $U^{(\gamma)}$ of (1.5).

The invariance of (1.3) with respect to local scale transformations

$$(d\xi, d\eta) \longrightarrow (d\xi', d\eta') = (h(\xi)\, d\xi,\ k(\eta)\, d\eta)\ , \quad h, k > 0\ ,$$

allows us, choosing $h(\xi) = \|\partial_\xi \vec{q}\|$, $k(\eta) = \|\partial_\eta \vec{q}\|$, to obtain the sine-Gordon theory itself: in the new, normalized co-ordinates, we have

$$\|\partial_{\xi'} \vec{q}\| = \|\partial_{\eta'} \vec{q}\| = 1\ , \tag{1.10}$$

and (1.8) and (1.9) reduce to the sine-Gordon equation and to its well-known Lax pair, respectively [written in the fundamental representation of O(3) rather than SU(2)]. The sine-Gordon theory is known to be a Hamiltonian system, completely integrable by the inverse scattering method[11]. It has an infinite series of higher local conservation laws starting with energy momentum conservation:

$$\partial_\eta \tfrac{1}{2} (\partial_\xi \alpha)^2 = \partial_\xi \cos\alpha \tag{1.11.1}$$

$$\partial_\eta \left\{ \tfrac{1}{2} (\partial_\xi^2 \alpha)^2 - \tfrac{1}{8} (\partial_\xi \alpha)^4 \right\} = \partial_\xi \left\{ -\tfrac{1}{2} (\partial_\xi \alpha)^2 \cos\alpha \right\} \tag{1.11.2}$$

etc.

The continuity equations (1.11), together with those obtained by interchanging ξ and η, provide a complete set of action variables for the sine-Gordon system. Using (1.7), they can be re-expressed in terms of the $\vec{q}$ field and its derivatives[7]:

$$\partial_\eta \left\{ \tfrac{1}{2} (\partial_\xi^2 \vec{q})^2 \right\} = \partial_\xi (\partial_\xi \vec{q} \cdot \partial_\eta \vec{q}) \tag{1.12.1}$$

$$\partial_\eta \left\{ \tfrac{1}{2} (\partial_\xi^3 \vec{q})^2 - \tfrac{5}{8} (\partial_\xi^2 \vec{q})^4 \right\} = \partial_\xi \left\{ -\tfrac{1}{2} (\partial_\xi^2 \vec{q})^2 (\partial_\xi \vec{q} \, \partial_\eta \vec{q}) \right\} \tag{1.12.2}$$

etc.

(Of course, the restriction to the normalized co-ordinates (1.10) can be eliminated by replacing

$$\partial_\xi \longrightarrow \| \partial_\xi \vec{q} \|^{-1} \partial_\xi \ , \quad \partial_\eta \longrightarrow \| \partial_\eta \vec{q} \|^{-1} \partial_\eta \ .)$$

Thus, the Lax pair (1.9) produces at the same time an infinite set of local conservation laws for the σ model itself. However, it is not known whether one can construct a complete set of action variables for the σ model out of both series of non-local (1.6) and local (1.12) conservation laws.

It has turned out[1)-3)] that the properties

(1.5) - Lax pair for the σ model

(1.6) - higher non-local conservation laws for the σ model

(1.8) - reduced system

(1.9) - Lax pair for the reduced system

(1.11)-(1.12) - higher local conservation laws for both the reduced system and the σ model

are not restricted to the S^2 model. They can be extended to a considerably large class of appropriately generalized σ models, and this will be explained in Chapters 2 - 6.

2. CONSTRUCTION OF NON-LINEAR σ MODELS ON RIEMANNIAN HOMOGENEOUS SPACES

Since the construction of generalized σ models has been treated extensively in the original papers[1)-4)] as well as in reviews[5),6)], I shall be rather brief. Let us start with a Lie group G with Lie algebra $\mathcal{G}$ and a closed subgroup H of G with Lie algebra $\mathcal{H}$. Denoting the complement of $\mathcal{H}$ in $\mathcal{G}$ by m, we have

$$\begin{gathered} \mathcal{G} = \mathcal{H} \oplus m \\ [\mathcal{H}, \mathcal{H}] \subset \mathcal{H} \ , \quad [\mathcal{H}, m] \subset m \ . \end{gathered} \tag{2.1}$$

We require the existence of an inner product (,) on $\mathcal{G}$ which is invariant under the adjoint representation of G and positive definite on m and which extends to the tangent bundle of the coset space G/H so that the latter becomes a Riemannian homogeneous space.

The non-linear σ model on G/H is defined in terms of a G valued field $g(x^0,x^1)$ with the action

$$S = \frac{1}{2}\int d^2x\,(D_\mu g, D^\mu g) \tag{2.2}$$

where the covariant derivative $D_\mu g$ of g with respect to the "gauge group" H is given by decomposing (cf. 2.1) the tangent vector $g^{-1}\partial_\mu g$ into its vertical part $A_\mu \in \mathcal{H}$ (the gauge field) and its horizontal part $g^{-1} D_\mu g \in m$ (the left translated covariant derivative):

$$\begin{array}{ccccc} g^{-1}\partial_\mu g & = & A_\mu & + & g^{-1}D_\mu g \\ \cap & & \cap & & \cap \\ \mathcal{G} & & \mathcal{H} & & m \end{array} \tag{2.3}$$

The action (2.2) is invariant under Lorentz and conformal transformations, under local H valued gauge transformations

$$g(x) \longrightarrow g(x)\,h(x) \quad , \quad h(x) \in H \;,$$

and under global G valued transformations

$$g(x) \longrightarrow g_0\, g(x) \quad , \quad g_0 \in G \text{ constant.}$$

The continuity equation corresponding to the global G invariance is

$$\partial_\mu j^\mu = 0 \quad , \quad j_\mu = -D_\mu g\, g^{-1} . \tag{2.4}$$

At the same time, (2.4) is the field equation corresponding to (2.2); usually it is written in the form

$$D_\mu D^\mu g - D_\mu g\, g^{-1} D^\mu g = 0 . \tag{2.5}$$

The simplest non-trivial example for this general prescription is given by

$$G = O(3)\,,\quad H = O(2)\,,\quad G/H = S^2 . \tag{2.6}$$

Then $$\mathcal{G} = \{ x \mid x \text{ real antisymmetric } 3\times 3 \text{ matrix} \}$$
$$\mathcal{H} = \left\{ \begin{pmatrix} 0 & 0 & 0 \\ 0 & \multicolumn{2}{c}{x} \\ 0 & & \end{pmatrix} \;\middle|\; x \text{ real antisymmetric } 2\times 2 \text{ matrix} \right\}$$
$$\mathcal{m} = \left\{ \begin{pmatrix} 0 & -z^T \\ z & \begin{smallmatrix} 0 & 0 \\ 0 & 0 \end{smallmatrix} \end{pmatrix} \;\middle|\; z = \begin{pmatrix} z_1 \\ z_2 \end{pmatrix} \in \mathbb{R}^2 \right\} .$$

Writing $g = (\vec{q},Y) = (\vec{q},\vec{y}^{(1)},\vec{y}^{(2)})$ for the field, we require $g \in O(3)$, i.e.,

$$g^T g = \mathbb{1}_3 , \quad \vec{q}^{\,2} = 1 , \quad Y^T Y = \mathbb{1}_2 , \quad \vec{q}^{\,T} Y = 0 . \tag{2.8}$$

The gauge group H acts on g as

$$(\vec{q},Y) \to (\vec{q}, Yh) \quad \text{with } h(x^0,x^1) \in O(2) . \tag{2.9}$$

According to (2.3),

$$A_\mu = \begin{pmatrix} 0 & 0 & 0 \\ 0 & \multicolumn{2}{c}{Y^T \partial_\mu Y} \\ 0 & & \end{pmatrix} , \quad D_\mu(\vec{q},Y) = (\partial_\mu \vec{q}, \partial_\mu Y - YY^T\partial_\mu Y) = (\partial_\mu \vec{q}, D_\mu Y) . \tag{2.10}$$

Inserting (2.10) into (2.2), (2.5) and (2.4) [the inner product is the Cartan-Killing form of O(3)], we find the action

$$S = \int d^2x \, (\partial_\mu \vec{q} \, \partial^\mu \vec{q}) = \int d^2x \, \mathrm{tr}\left(D_\mu Y (D^\mu Y)^T\right) , \tag{1.2}$$

the field equations

$$\Box \vec{q} + (\partial_\mu \vec{q} \, \partial^\mu \vec{q}) \vec{q} = 0 \quad \text{and} \quad D_\mu D^\mu Y + Y (D_\mu Y)^T D^\mu Y = 0 , \tag{1.3}$$

and the current

$$j_\mu = \vec{q} (\partial_\mu q)^T - (\partial_\mu q) \vec{q}^{\,T} = Y (D_\mu Y)^T - (D_\mu Y) Y^T . \tag{1.4}$$

Thus we have rederived the S^2 non-linear σ model; the generalization to $S^{N-1} = O(N)/O(N-1)$ is obvious. More complicated examples, like the complex Grassmannian $SU(p+q)/S[U(p) \times U(q)]$, can be found in the literature cited above.

As the reader will have noticed, the main ingredient in the definition of the NLSM is not the global symmetry group G, but the homogeneous space G/H. Therefore, we prefer the terminology "$S^{N-1}\sigma$ model" rather than "$O(N)\sigma$ model". For $N > 3$, there are various different NLSM with global $O(N)$ symmetry corresponding to the various closed subgroups of $O(N)$.

As a side remark, observe that we obtain two different formulations of the S^{N-1} model: the first (the usual one) involves the field vector $\vec{q}$ and is gauge invariant, whereas the second one involves the Y field and has a (for $N > 3$) non-Abelian $O(N-1)$ gauge symmetry. Classically, this simply amounts to different parametrizations of S^{N-1}. For the corresponding quantum field theories, however, the consequences are not clear: for example, the $1/N$ expansion, working smoothly in terms of $\vec{q}$, in terms of Y immediately leads to the problem of summing planar Feynman graphs.

3. LAX REPRESENTATION, NON-LOCAL CONSERVATION LAWS AND SYMMETRIC SPACES

At present, there exists no systematic way of deciding whether a given two dimensional model has a Lax representation or not. We can, however, take Eq. (1.5) as a guide and try to generalize it. So let us look at the ansatz

$$\begin{aligned} \partial_\xi U^{(\gamma)} &= (1-\gamma^{-1})\, U^{(\gamma)} j_\xi \\ \partial_\eta U^{(\gamma)} &= (1-\gamma)\, U^{(\gamma)} j_\eta \end{aligned} \tag{3.1}$$

where $j_\mu = -D_\mu g g^{-1} \in \mathcal{G}$, $U^{(\gamma)} \in G$, $\gamma \in \mathbb{R}$. (3.1) is a system of linear differential equations on the group G. Given the components j_ξ and j_η of the conserved G current, (3.1) determines a one parameter family of G valued transformations $U^{(\gamma)}$ (up to normalization) provided both equations are compatible. Using current conservation, i.e., the field equation, the compatibility condition for (3.1)

$$\partial_\xi (1-\gamma) j_\eta - \partial_\eta (1-\gamma^{-1}) j_\xi + [(1-\gamma^{-1}) j_\xi, (1-\gamma) j_\eta] = 0 \tag{3.2}$$

takes the form

$$\partial_\xi j_\eta - \partial_\eta j_\xi + 2[j_\xi, j_\eta] = 0 . \tag{3.3}$$

Inserting $j_\mu = -D_\mu g g^{-1}$ and observing that (2.3) implies

$$F_{\xi\eta} = \partial_\xi A_\eta - \partial_\eta A_\xi + [A_\xi, A_\eta] = [g^{-1}D_\eta g, g^{-1}D_\xi g]_{\mathcal{H}} , \tag{3.4}$$

(3.3) can be written as

$$[g^{-1}D_\xi g, g^{-1}D_\eta g]_m = 0 . \tag{3.5}$$

(Subscripts $\mathcal{H}$, m, etc., always denote projection to the respective subspaces.) Because of $g^{-1}D_\mu g \in m$, this gives the additional condition

$$[m, m] \subset \mathcal{H} . \tag{3.6}$$

(3.6) is the (infinitesimal) defining property of a certain subclass of homogeneous spaces, the so-called symmetric spaces[12)].

Before explaining this in more detail, let me continue the discussion of (3.1). Whenever we start from a homogeneous space G/H satisfying (3.6), the linear equations (3.1) have the field equation of the NLSM on G/H as their compatibility condition. Hence (3.1) provides a Lax representation for the σ model the status of which is the same as for the S^2 case: for boundary conditions like $j_\mu \to 0$ as $|x^1| \to \infty$, we do not know how to solve (2.5) via (3.1) in the sense of the inverse scattering method. Nevertheless, (3.1) turns out to be the basis of all integrability properties of the NLSM; therefore, I shall concentrate in the following on the class of NLSM on symmetric spaces.

As for the S^2 case, we can derive an infinite series of non-local conservation laws from (3.1), proceeding as follows. Write

$$U^{(\gamma)} = \exp V^{(\gamma)} \tag{3.7}$$

where $v^{(\gamma)} \in \mathcal{G}$. The Lie algebra $\mathcal{G}$ being a linear space, it makes sense to perform a power series expansion of $v^{(\gamma)}$ around $\gamma = 1$:

$$V^{(\gamma)} = \sum_{k=1}^{\infty} \epsilon^k V^{(k)} \tag{3.8}$$

where $\epsilon = 1 - \gamma$ and the normalization $U^{(1)} = \mathbb{1}$ has been chosen. Insert (3.7) and (3.8) into (3.1):

$$\begin{aligned} (\varepsilon-1) \exp\Big\{-\sum_{k=1}^{\infty} \varepsilon^k V^{(k)}\Big\} \partial_\xi \exp\Big\{\sum_{k=1}^{\infty} \varepsilon^k V^{(k)}\Big\} &= \varepsilon j_\xi \\ \exp\Big\{-\sum_{k=1}^{\infty} \varepsilon^k V^{(k)}\Big\} \partial_\eta \exp\Big\{\sum_{k=1}^{\infty} \varepsilon^k V^{(k)}\Big\} &= \varepsilon j_\eta . \end{aligned} \tag{3.9}$$

Using the formula for the derivative of the exponential map[12)]

$$\exp(-X)\,\partial_\mu(\exp X) = \sum_{n=0}^{\infty} \frac{(-1)^n}{(n+1)!} \underbrace{[X, \ldots, [X, \partial_\mu X]\ldots]}_{n \text{ brackets}} \tag{3.10}$$

and collecting the terms with the same power of ϵ in (3.9), one can recursively determine the $\partial_\xi v^{(k)}$, $\partial_\eta v^{(k)}$, $v^{(k)}$ (non-local!) to obtain the $\mathcal{G}$ valued currents entering the conservation laws

$$\partial_\eta(\partial_\xi V^{(k)}) - \partial_\xi(\partial_\eta V^{(k)}) = 0 \qquad k = 1\ldots\infty\,. \tag{3.11}$$

Explicitly:

$$k=1 \qquad \partial_\eta j_\xi + \partial_\xi j_\eta = 0 \tag{3.12.1}$$

$$k=2 \qquad \partial_\eta \left\{ \tfrac{1}{2}\left[\int_{-\infty}^{\xi} d\xi' \, j_\xi(\xi',\eta), j_\xi(\xi,\eta)\right] - j_\xi(\xi,\eta)\right\}$$
$$+ \partial_\xi \left\{ \tfrac{1}{2}\left[\int_{-\infty}^{\xi} d\xi' \, j_\xi(\xi',\eta), j_\eta(\xi,\eta)\right]\right\} = 0 \tag{3.12.2}$$

etc.

Let us now look closer at condition (3.6). Not all homogeneous spaces fulfill it: counterexamples are the real and complex Stiefel manifolds

$$\left.\begin{matrix} SO(N)/SO(M) \\ SU(N)/SU(M) \end{matrix}\right\} \; 2 \le M \le N-2$$

or flag manifolds like

$$U(3)/U(1)\times U(1)\times U(1) \quad .$$

However, decomposing the respective Lie algebras, as done in (2.7), the reader can easily convince himself that the following homogeneous spaces match with (3.6): The real and complex Grassmann manifolds

$$SO(p+q)/SO(p)\times SO(q)$$
$$SU(p+q)/S(U(p)\times U(q))$$

containing the spheres S^{N-1} discussed in Chapter 2 as special cases,or the spaces

$$SU(N)/SO(N)$$
$$SO(2N)/U(N)$$

etc.

Actually, E. Cartan was able to give a complete list of the symmetric spaces. This list can be found in Ref. 12) p. 516 ff; it contains

- several series of compact spaces based on classical groups
- several compact spaces based on exceptional groups
- non-compact ("dual") versions of both the classical and exceptional spaces.

Cartan's classification exhibits yet another important type of symmetric space, namely, the Lie groups themselves. In fact, observe that any Lie group G with Lie algebra $\mathcal{G}$ can be represented as $G \simeq G \times G/\Delta G$ where $\Delta G = \{(g,g) | g \in G\}$ is the diagonal of $G \times G$, the diffeomorphism being induced by the map

$$\Gamma : G \times G \longrightarrow G \quad , \quad \Gamma(g_1, g_2) = g_1 g_2^{-1}$$

with

$$\Gamma((g_1, g_2)\cdot(g,g)) = \Gamma(g_1 g, g_2 g) = \Gamma(g_1, g_2) \; .$$

Then

$$\mathcal{G} \oplus \mathcal{G} = \mathcal{H} \oplus m$$

where

$$\mathcal{H} = \{(a,a) \,|\, a \in \mathcal{G}\}$$

$$m = \{(a,-a) \,|\, a \in \mathcal{G}\}$$

and obviously

$$[m, m] \subset \mathcal{H} \; .$$

The corresponding NLSM can be formulated either in terms of a $G \times G$ valued field with gauge group ΔG or in terms of a G valued field without gauge group[1),2)]. In both cases, the model has a global $G \times G$ symmetry motivating the name "chiral field model". In the Russian literature[13)], the chiral models are called "principal models". The reason for this is the following: one can show[2)] that the solution space of the principal model on a Lie group G contains the solution spaces of the NLSM on all spaces G/H, H being any subgroup of G so that G/H is symmetric, as subspaces. Thus a complete description of the principal field dynamics implies a complete description of the dynamics of all G/H models where G/H is symmetric.

4. SOME MATHEMATICAL TOOLS

In the first lecture, dealing with the general construction of σ models, the determination of the Lax representation and the computation of the non-local conservation laws, the main role was played by the $\mathcal{G}$ valued currents

$$j_\mu = - D_\mu g \, g^{-1} \; .$$

In the second lecture, dealing with the generalization of the reduction procedure (1.8) and (1.9) and the computation of local conservation laws for NLSM on symmetric spaces, this role will be taken over by the quantities

$$A_\mu \in \mathcal{H} \quad \text{and} \quad k_\mu = g^{-1} D_\mu g \in m$$

defined by (2.3) because these directly reveal the decomposition of $\mathcal{G}$ into its vertical and horizontal part.

For symmetric spaces, the field equation (2.5) takes the form

$$D_\eta D_\xi g - D_\eta g\, g^{-1} D_\xi g = 0 \quad , \text{ in terms of } k_\mu : \; D_\eta k_\xi = 0 \tag{4.1}$$

or, equivalently,

$$D_\xi D_\eta g - D_\xi g\, g^{-1} D_\eta g = 0 \quad , \text{ in terms of } k_\mu : \; D_\xi k_\eta = 0 \, . \tag{4.2}$$

Here,

$$D_\mu k_\nu = \partial_\mu k_\nu - [k_\nu , A_\mu] \in m \, .$$

Furthermore, (3.4) and (3.6) give the equation

$$F_{\xi\eta} = \partial_\xi A_\eta - \partial_\eta A_\xi + [A_\xi , A_\eta] = [k_\eta , k_\xi] \, . \tag{4.3}$$

The fields A_μ and k_μ will enter Chapters 5 and 6 mainly through Eqs. (4.1) - (4.3).

In the following, we shall need a refinement of the decomposition $\mathcal{G} = \mathcal{H} \oplus m$ [12].

- let $\mathcal{A}$ denote any maximal Abelian subspace of m,
- define ℓ to be that subalgebra of $\mathcal{H}$ the elements of which commute with all elements of $\mathcal{A}$ (the centralizer of $\mathcal{A}$ in $\mathcal{H}$),
- define n to be the orthogonal complement of $\mathcal{A}$ in m,
- define p to be the orthogonal complement of ℓ in $\mathcal{H}$

Then we have the orthogonal decompositions

$$\mathcal{H} = \ell \oplus p \; , \quad m = \mathcal{A} \oplus n \; , \quad \mathcal{G} = \ell \oplus p \oplus \mathcal{A} \oplus n \quad . \tag{4.4}$$

Frequently, we shall use the corresponding decompositions of A_μ and k_μ:

$$A_\mu = (A_\mu)_\ell + (A_\mu)_p \ , \qquad K_\mu = (K_\mu)_{\mathcal{A}} + (K_\mu)_n \ . \tag{4.5}$$

The various subspaces obey the commutation relations

$$\begin{array}{llll} [\mathcal{A},\mathcal{A}] = \{0\} & [\mathcal{A},\ell] = \{0\} & [\ell,\ell] \subset \ell & [p,p] \subset \mathcal{H} \\ [\mathcal{A},p] \subset n & & [\ell,p] \subset p & [p,n] \subset m \\ [\mathcal{A},n] \subset p & & [\ell,n] \subset n & [n,n] \subset \mathcal{H} \ . \end{array} \tag{4.6}$$

The dimension of $\mathcal{A}$ is called the rank of the symmetric space G/H. Remember that Lie groups themselves are symmetric spaces: in that case, the above notions of maximal Abelian subspace and rank of the symmetric space coincide with the usual notions of Cartan subalgebra and rank of the group.

For the example $G/H = S^2$, we have

$$\begin{array}{l} \mathcal{A} = \left\{ \begin{pmatrix} 0 & -x & 0 \\ x & 0 & 0 \\ 0 & 0 & 0 \end{pmatrix} \middle| \, x \in \mathbb{R} \right\} , \quad n = \left\{ \begin{pmatrix} 0 & 0 & -x \\ 0 & 0 & 0 \\ x & 0 & 0 \end{pmatrix} \middle| \, x \in \mathbb{R} \right\} , \\ \ell = \{0\} \qquad\qquad , \quad p = \mathcal{H} \ , \\ k_\mu = \begin{pmatrix} 0 & \vec{q}^{\,T} D_\mu Y \\ Y^T \partial_\mu \vec{q} & \begin{matrix} 0 & 0 \\ 0 & 0 \end{matrix} \end{pmatrix} \quad , \quad \mathrm{rank}(S^2) = 1 \ . \end{array} \tag{4.7}$$

For more complicated examples consult Ref. 3).

As a second tool we shall use the polar decomposition theorem[12),3)] for symmetric spaces. In its infinitesimal version, it states that each element $Z \in m$ can be written in the form

$$Z = h x h^{-1} \quad \text{where } h \in H \ , \ x \in \mathcal{A} \ . \tag{4.8}$$

Essentially [i.e., up to discrete Weyl group transformations, cf., Ref. 3)], x is unique and h is determined up to transformations taking values in the centralizer L of $\mathcal{A}$ in H:

$$L = \{ h \in H \mid h x h^{-1} = x \text{ for all } x \in \mathcal{A} \} \quad . \tag{4.9}$$

Observe that the Lie algebra of L coincides with ℓ defined above.

For the S^2 example, let

$$Z = \begin{pmatrix} 0 & -z_1 & -z_2 \\ z_1 & 0 & 0 \\ z_2 & 0 & 0 \end{pmatrix} \in m, \quad h = \begin{pmatrix} 0 & 0 & 0 \\ 0 & \cos\alpha & -\sin\alpha \\ 0 & \sin\alpha & \cos\alpha \end{pmatrix} \in H, \quad x = \begin{pmatrix} 0 & -r & 0 \\ r & 0 & 0 \\ 0 & 0 & 0 \end{pmatrix} \in \mathcal{A}, \; r > 0. \tag{4.10}$$

Then (4.8) reduces to the usual polar co-ordinates of $\mathbb{R}^2$:

$$\begin{pmatrix} z_1 \\ z_2 \end{pmatrix} = \begin{pmatrix} r\cos\alpha \\ r\sin\alpha \end{pmatrix} \tag{4.11}$$

(The global version of the polar decomposition theorem yields the usual polar co-ordinates on $\exp m \simeq S^2$.)

5. REDUCTION GAUGE AND REDUCED SYSTEMS

To construct the analogue of (1.8) and (1.9), the reduced system, for the general situation, we shall use the fact that for the NLSM on a symmetric space there exists a distinguished gauge which essentially completely removes the freedom to perform H valued gauge transformations. This gauge is constructed in two steps.

First, applying the polar decomposition (4.8) to $Z = k_\xi \in m$ and taking its gauge transformation behaviour

$$k_\xi = g^{-1} D_\xi g \longrightarrow h^{-1} k_\xi h$$

into account, we see that we can always choose a gauge so that

$$k_\xi \in \mathcal{A}. \tag{5.1}$$

This is a <u>local</u> gauge still leaving us with the freedom to perform residual gauge transformations with values in L (4.9). Inserting the decomposition $A_\eta = (A_\eta)_\ell + (A_\eta)_p$ into Eq. (4.1) and using $[\mathcal{A},\ell] = \{0\}$ and $[\mathcal{A},p] \subset n$, we find the equations

$$\begin{aligned} 0 &= \partial_\eta k_\xi && \in \mathcal{A} \\ 0 &= [k_\xi, (A_\eta)_p] && \in n \quad . \end{aligned} \tag{5.2}$$

Assuming that k_ξ fulfills certain regularity conditions [cf., Ref. 3)], then we can invert the commutator in the second equation to get

$$(A_\eta)_p = 0 \,. \tag{5.3}$$

Second, we may use the residual gauge transformations to gauge away the ℓ part of A_ξ and A_η. In fact, the differential equations

$$\begin{aligned} \partial_\xi h^R + (A_\xi)_\ell h^R &= 0 \\ \partial_\eta h^R + (A_\eta)_\ell h^R &= 0 \end{aligned} \tag{5.4}$$

have the equation

$$\partial_\xi (A_\eta)_\ell - \partial_\eta (A_\xi)_\ell + [(A_\xi)_\ell, (A_\eta)_\ell] = 0 \tag{5.5}$$

as their compatibility condition. However, (5.5) is precisely the ℓ part of Eq. (4.3):

$$\begin{aligned} 0 &= [k_\eta, k_\xi]_\ell \\ &= (\partial_\xi A_\eta)_\ell - (\partial_\eta A_\xi)_\ell + [(A_\xi)_\ell + (A_\xi)_p, (A_\eta)_\ell]_\ell \\ &= \partial_\xi (A_\eta)_\ell - \partial_\eta (A_\xi)_\ell + [(A_\xi)_\ell, (A_\eta)_\ell] \end{aligned}$$

due to $[\mathcal{A}, n] \subset p$, (5.3), $[\ell, p] \subset p$, $[\ell, \ell] \subset \ell$. Hence, (5.4) defines an L valued gauge transformation h^R which exactly cancels $(A_\xi)_\ell$ and $(A_\eta)_\ell$. We call this gauge, in which

$$K_\xi^R \in \mathcal{A} \,, \quad (A_\xi^R)_\ell = (A_\eta^R)_\ell = (A_\eta^R)_p = 0 \tag{5.6}$$

the reduction gauge. All quantities in the reduction gauge will be distinguished by an upper index R. For L non-trivial, it is a non-local gauge due to (5.4) leaving us finally with the freedom to perform <u>constant</u>, L valued gauge transformations.

In the reduction gauge, Eqs. (4.1) - (4.3) take the following form for the remaining non-zero degrees of freedom $k_\xi^R \in \mathcal{A}, k_\eta^R \in m$, $A_\xi^R \in p$:

$$\partial_\eta k_\xi^R = 0 \tag{4.1^R}$$

$$\partial_\xi k_\eta^R = [k_\eta^R, A_\xi^R] \tag{4.2^R}$$

$$\partial_\eta A_\xi^R = [k_\xi^R, (k_\eta^R)_n] \,. \tag{4.3^R}$$

For k_ξ^R regular (cf. above), we can solve (4.3^R) for $(k_\eta^R)_n$:

$$(k_\eta^R)_n = ad(k_\xi^R)^{-1}\, \partial_\eta A_\xi^R . \tag{5.7}$$

Here we use the notation $ad(x)Y = [x,Y]$ for $x \in \mathcal{A}$, $Y \in n\ (p)$, the inverse operation $ad(x)^{-1}$ for x regular being a linear transformation $p \to n\ (n \to p)$. Introducing a field $\Phi \in n$ by

$$\Phi = ad(k_\xi^R)^{-1} A_\xi^R ,$$

we have

$$k_\eta^R = \partial_\eta \Phi \tag{5.8}$$

by (4.1^R). Decomposing (4.2) into its $\mathcal{A}$ and its n part and inserting (5.8), we finally arrive at the reduced system defined by the following set of non-linear partial differential equations

$$\partial_\eta k_\xi^R = 0 \tag{5.9.1}$$

$$\partial_\xi (k_\eta^R)_{\mathcal{A}} = \tfrac{1}{2}\, \partial_\eta [\Phi, [k_\xi^R, \Phi]]_{\mathcal{A}} \tag{5.9.2}$$

$$\Box \Phi = [\partial_\eta \Phi, [k_\xi^R, \Phi]]_n + [(k_\eta^R)_{\mathcal{A}}, [k_\xi^R, \Phi]] \tag{5.9.3}$$

for the variables

$$k_\xi^R \in \mathcal{A} , \quad (k_\eta^R)_{\mathcal{A}} \in \mathcal{A} , \quad \Phi \in n . \tag{5.10}$$

Equations (5.9) are the compatibility conditions for the Lax pair

$$\begin{aligned} \partial_\xi g^{(\gamma)} &= g^{(\gamma)}\left([k_\xi^R, \Phi] + \gamma^{-1} k_\xi^R \right) \\ \partial_\eta g^{(\gamma)} &= \gamma\, g^{(\gamma)}\left((k_\eta^R)_{\mathcal{A}} + \partial_\eta \Phi \right) \end{aligned} \tag{5.11}$$

which is closely connected to the Lax pair (3.1) for the σ model itself: in fact, define

$$g^{(\gamma)} = u^{(\gamma)} g . \tag{5.12}$$

(3.1) then implies

$$\partial_\xi g^{(\gamma)} = g^{(\gamma)}(A_\xi + \gamma^{-1} k_\xi)$$
$$\partial_\eta g^{(\gamma)} = g^{(\gamma)}(A_\eta + \gamma\, k_\eta) \quad , \tag{5.13}$$

and using the reduction gauge conditions (5.6) and the defintion of Φ, we arrive at (5.11).

Actually, this shows that in general the reduced system (5.9) and the NLSM are related by the principle of "gauge equivalence of two dimensional integrable models"[10),13)]. First, by the (G valued) "gauge transfomation" (5.12) we go over from the Lax pair (3.1) to the new Lax pair (5.13) - the compatibility is preserved because it is a flatness condition, cf., (3.2). Second, the high degree of gauge ambiguity in (5.12) is eliminated by choosing the reduction gauge, thus arriving at (5.11); this step motivates the name "reduced systems". Finally, the compatibility equations of (5.11) define the "dynamics" of the new model.

The experience with the sine-Gordon case leads us to hope that for suitable boundary conditions such as

$$\left.\begin{array}{lll} \Phi \to 0 & , \quad k_\xi^R & \longrightarrow \text{const.}_1 \in \mathcal{A} \\ \partial_\eta \Phi \to 0 & , \quad (k_\eta^R)_{\mathcal{A}} & \longrightarrow \text{const.}_2 \in \mathcal{A} \end{array}\right\} \text{ as } |x^1| \to \infty \text{ , cf. Ref. 17)},$$

the reduced systems can be solved by the inverse scattering method based on (5.11). As a by-product of this, one could also determine the original NLSM field $g = g^{(1)}$ from the "wave function" $g^{(\gamma)}$ of the "scattering problem" (5.11). However, the above boundary conditions do not match with $j_\mu \to 0$, so this is not a way of overcoming the difficulties with the NLSM itself.

To consider (5.9) and (5.11) as the reduced system generalizing (1.8) and (1.9) is justified by the fact that in the well-known special cases S^{N-1} [14)] and $\mathbb{C}P^{N-1}$ [15)] it reproduces the former results. Actually, up to slight modifications, the procedure described above is the direct generalization of the approach used in Ref. 15).

Let me now show that for S^2, we rederive our starting point (1.8) and (1.9). According to (4.7) we write

$$k_\xi^R = \begin{pmatrix} 0 & -k & 0 \\ k & 0 & 0 \\ 0 & 0 & 0 \end{pmatrix} \quad , \quad (k_\eta^R)_{\mathcal{A}} = \begin{pmatrix} 0 & -\lambda & 0 \\ \lambda & 0 & 0 \\ 0 & 0 & 0 \end{pmatrix} \quad , \quad \Phi = \begin{pmatrix} 0 & 0 & -\varphi \\ 0 & 0 & 0 \\ \varphi & 0 & 0 \end{pmatrix} \quad . \tag{5.14}$$

Inserting (5.14) into (5.9), we get

$$\begin{aligned} \partial_\eta k &= 0 \\ \partial_\xi \lambda &= \tfrac{1}{2}\partial_\eta (k\varphi^2) \\ \Box\varphi + k\lambda\varphi &= 0 . \end{aligned} \tag{5.15}$$

Choosing now $\lambda = \delta\cos\alpha$, $\partial_\eta \phi = \delta\sin\alpha$ $(\delta > 0)$, (5.15) takes the form

$$\begin{aligned} \partial_\eta k &= 0 \\ \partial_\xi \delta &= 0 \\ \Box\alpha + k\delta\sin\alpha &= 0 , \end{aligned} \tag{1.8}$$

and (5.11) yields the Lax pair (1.9). In normalized co-ordinates (1.10) we even have $k = \delta = 1$, and the reduced system is the sine-Gordon equation itself. In addition, we note that the gauge condition (5.1)

$$k_\xi = g^{-1} D_\xi g = \begin{pmatrix} 0 & -1 & 0 \\ 1 & 0 & 0 \\ 0 & 0 & 0 \end{pmatrix}$$

implies

$$D_\xi g = (\vec{q}, \vec{y}^{(1)}, \vec{y}^{(2)}) \begin{pmatrix} 0 & -1 & 0 \\ 1 & 0 & 0 \\ 0 & 0 & 0 \end{pmatrix} = (\vec{y}^{(1)}, -\vec{q}, 0) ,$$

i.e.,

$$g = (\vec{q}, \partial_\xi \vec{q}, \vec{y}^{(2)}) \quad \text{and} \quad \vec{y}^{(2)} = \pm \frac{1}{\sin\alpha} (\partial_\eta \vec{q} - \cos\alpha\, \partial_\xi \vec{q}). \tag{5.16}$$

This is the basis of $\mathbb{R}^3$ used originally in Ref. 7) to reduce the S^2 σ model to the sine-Gordon theory. More complicated examples are discussed in Ref. 16).

Unfortunately, up to now we do not know how to derive (5.9) from a Lagrangian density, except for the special cases S^2, S^3 and $\mathbb{C}P^2$ which are discussed in Refs. 7) and 15). These examples indicate that the variables (5.10) might not be the appropriate ones to find a Lagrangian density for (5.9) and to bring the reduced systems into the status of a well-behaved Lagrangian field theory. However, the following general characteristical properties should hold in any, perhaps more suitable parametrization:

i) The number of independent variables is dim(G/H) + rank(G/H). Two more degrees of freedom could be eliminated by the co-ordinate normalization (1.10), choosing $h(\xi) = (D_\xi g, D_\xi g)^{1/2}$, $k(\eta) = (D_\eta g, D_\eta g)^{1/2}$, but it is not clear whether this is helpful if rank(G/H) > 1.

ii) The definition of A_μ and k_μ shows that the variables (5.10) are globally G invariant.

iii) The local gauge freedom has been eliminated, leaving us with a global L symmetry:

$$\Phi \to h^{-1} \Phi h \qquad h \in L \text{ constant.}$$

iv) The equations (5.9) are Lorentz invariant, the couple $(k_\xi^R, (k_\eta^R)_{\mathcal{A}})$ transforming like a vector and Φ like a scalar.

6. LOCAL CONSERVATION LAWS FOR NLSM AND REDUCED SYSTEMS

Our last topic is to find the analogues of the local conservation laws (1.11) and (1.12) in the general situation. Guided by our experience from various special models, we start from the fact that the field $g^{(\gamma)}$ (5.12) can locally be decomposed uniquely as follows:

$$g^{(\gamma)} = \exp a^{(\gamma)} \exp \ell^{(\gamma)} \exp \omega^{(\gamma)} \tag{6.1}$$

where $a^{(\gamma)} \in \mathcal{A}$, $\ell^{(\gamma)} \in \ell$, $\omega^{(\gamma)} \in n \oplus p$. We claim that the function $a^{(\gamma)}$ is the generating functional for higher, gauge invariant local conservation laws. First let me sketch how to compute $a^{(\gamma)}$, then the questions of gauge invariance and locality will be discussed.

In the gauge (5.1) which will be used throughout this Chapter, it turns out that the functions $a^{(\gamma)}$ and $\omega^{(\gamma)}$ have power series expansions of the form

$$a^{(\gamma)} = \sum_{k=-1}^{\infty} \gamma^k a^{(k)} \quad , \qquad \omega^{(\gamma)} = \sum_{k=1}^{\infty} \gamma^k \omega^{(k)} \tag{6.2}$$

with

$$a^{(k)} = 0 \text{ for } k \text{ even} \; ; \qquad \begin{array}{l} \omega^{(k)} \in n \quad \text{for } k \text{ odd}, \\ \omega^{(k)} \in p \quad \text{for } k \text{ even} \end{array} \tag{6.3}$$

Actually, (6.3) follows from a certain symmetry of $g^{(\gamma)}$ with respect to $\gamma \to -\gamma$[2),3)]. To compute the $a^{(k)}$ and $\omega^{(k)}$, we insert the decomposition (6.1) into the differential equations (5.13) and solve for the $\partial_\mu a^{(\gamma)}$. For the ξ equation, for example, this gives

$$\underbrace{\partial_\xi a^{(\gamma)}}_{\in \mathcal{A}} = \underbrace{e^{\omega^{(\gamma)}}(A_\xi + \gamma^{-1} k_\xi)e^{-\omega^{(\gamma)}} - (\partial_\xi e^{\omega^{(\gamma)}})e^{-\omega^{(\gamma)}}}_{\in \mathcal{A} \oplus n \oplus p \oplus \ell} + \underbrace{(\partial_\xi e^{-\ell^{(\gamma)}})e^{\ell^{(\gamma)}}}_{\in \ell} \tag{6.4}$$

$\mathcal{A}$ part: determines $a^{(\gamma)}$; $n \oplus p$ part: determines $\omega^{(\gamma)}$; ℓ part: decouples completely

Inserting (6.2) into (6.4) and the analogous η equation, and using once more (3.10), we can recursively compute the $\omega^{(k)}$ and $a^{(k)}$ as indicated in the scheme (6.4):

The $p \oplus n$ part determines the $\omega^{(k)}$ in terms of the $\omega^{(j)}$ $(1 \le j \le k-1)$, the $\mathcal{A}$ part determines the $a^{(k)}$ (or rather their derivatives) in terms of the $\omega^{(j)}$ $(1 \le j \le k)$, whereas, due to the commutation relations (4.6), the ℓ part decouples completely and need not be considered. In this way, we find the $\mathcal{A}$ valued currents entering the conservation laws

$$\partial_\eta(\partial_\xi a^{(2k+1)}) = \partial_\xi(\partial_\eta a^{(2k+1)}) \quad , \quad k = -1, \ldots, +\infty \tag{6.5}$$

Let us look at the first three orders of the recursion in more detail:

$$\gamma^{-1}: \quad \partial_\xi a^{(-1)} = k_\xi \tag{6.6}$$

$$\gamma^{0}: \qquad 0 = (A_\xi)_p + [\omega^{(1)}, k_\xi] \qquad \text{(p-part)}$$

$$\gamma^{1}: \begin{cases} \partial_\xi a^{(1)} = \left\{[\omega^{(1)}, (A_\xi)_p] + \tfrac{1}{2}[\omega^{(1)}, [\omega^{(1)}, k_\xi]]\right\}_{\mathcal{A}} & (\mathcal{A}\text{-part}) \\ 0 = \left\{-D_\xi \omega^{(1)} + [\omega^{(1)}, (A_\xi)_p] + [\omega^{(2)}, k_\xi] + \tfrac{1}{2}[\omega^{(1)}, [\omega^{(1)}, k_\xi]]\right\}_n & (\text{n-part}) \end{cases}$$

Here we have used the L covariant derivative

$$D_\xi \omega^{(1)} = \partial_\xi \omega^{(1)} - [\omega^{(1)}, (A_\xi)_\ell] \in n .$$

From (6.6) γ^{-1}, we directly get

$$\partial_\xi a^{(-1)} = k_\xi \quad . \tag{6.7}$$

The p part of (6.6) γ^0 gives

$$\omega^{(1)} = ad(k_\xi)^{-1}(A_\xi)_p \tag{6.8}$$

for k_ξ regular, cf., Chapter 5.

The $\mathcal{A}$ part of (6.6) γ^1 gives, using (6.8)

$$\partial_\xi a^{(1)} = \tfrac{1}{2}[ad(k_\xi)^{-1}(A_\xi)_p, (A_\xi)_p]_{\mathcal{A}} \quad . \tag{6.9}$$

The n part of (6.6) γ^1 serves to determine $\omega^{(2)}$:

$$\omega^{(2)} = ad(k_\xi)^{-1}\{-D_\xi\omega^{(1)} + [\omega^{(1)}, (A_\xi)_p] + \tfrac{1}{2}[\omega^{(1)}, [\omega^{(1)}, k_\xi]]\}_n .$$

Doing the same for the η equation, we find the first two conservation laws

$$\partial_\eta k_\xi = 0 \tag{6.10}$$

$$\partial_\eta \tfrac{1}{2}[ad(k_\xi)^{-1}(A_\xi)_p, (A_\xi)_p]_{\mathcal{A}} = \partial_\xi (k_\eta)_{\mathcal{A}} \quad . \tag{6.11}$$

The scheme indicated above holds to all orders. We determine

- the $\omega^{(2k)}$ from the n part of the γ^{2k-1} ξ equation
- the $\omega^{(2k+1)}$ from the p part of the γ^{2k} ξ equation
- the $\partial_\xi a^{(2k+1)}$ from the $\mathcal{A}$ part of the γ^{2K+1} ξ equation
- the $\partial_\eta a^{(2k+1)}$ from the $\mathcal{A}$ part of the γ^{2k+1} η equation .

The fact that the generating functional takes values in $\mathcal{A}$ implies that for a symmetric space of rank p, there exist p independent series of scalar local conservation laws. For the rank one spaces S^{N-1} and $\mathbb{C}P^{N-1}$, there is one series.

For the rank p spaces $SU(p+q)/S[U(p) \times U(q)]$, $p \leq q$ or for the principal model on a group of rank p, there are p independent series. Expressed in the variables (5.10), they are at the same time local conservation laws for the reduced models.

By the way, (6.10) and (6.11) coincide with the differential equations (5.9.1) and (5.9.2) for the reduced model, and in particular the derivation of (5.9.2) gives an impression of the sophisticated interplay of σ model dynamics (4.1) and (4.2) and symmetric space structure (4.3), (4.6) allowing for the conservation law (6.11).

I shall now demonstrate gauge invariance and locality for the example of the conservation law (6.11), referring the reader to Ref. 3) for the general case. Since we have used the gauge (5.1) to derive (6.11), we have to look at the behaviour of (6.11) under the residual, L valued gauge transformations preserving (5.1). Now $g \rightarrow gh$ for $h \in L$ implies

$$(k_\mu)_{\mathcal{A}} \longrightarrow (h^{-1} k_\mu h)_{\mathcal{A}} = (k_\mu)_{\mathcal{A}} \qquad (\text{observe } [\ell, \mathcal{A}] = \{0\}) \tag{6.12}$$

$$(k_\mu)_n \longrightarrow (h^{-1} k_\mu h)_n = h^{-1}(k_\mu)_n h \qquad (\text{observe } [\ell, n] \subset n) \tag{6.13}$$

and

$$A_\mu = (A_\mu)_\ell + (A_\mu)_p \longrightarrow (h^{-1}(A_\mu)_\ell h + h^{-1}\partial_\mu h) + h^{-1}(A_\mu)_p h$$

$$(\text{observe } [\ell, p] \subset p) ,$$

i.e.,

$$(A_\mu)_\ell \longrightarrow h^{-1}(A_\mu)_\ell h + h^{-1}\partial_\mu h \tag{6.14}$$

$$(A_\mu)_p \longrightarrow h^{-1}(A_\mu)_p h \quad . \tag{6.15}$$

Using (6.12) and (6.15) and again $[\ell, \mathcal{A}] = \{0\}$, we find that (6.11) is in fact invariant under residual gauge transformations. On the other hand, it is obviously local because we are working in the local gauge (5.1) and all manipulations leading from (6.4) to (6.11) are local ones.

Finally, let me show that for the S^2 example, (6.11) in fact reduces to the conservation law (1.12.1). From (5.16) and (2.10) we have

$$g = (\vec{q}, \partial_\xi \vec{q}, \vec{y}^{(2)}) , \quad (A_\xi)_p = \begin{pmatrix} 0 & 0 & 0 \\ 0 & & \\ 0 & \multicolumn{2}{c}{Y^T \partial_\xi Y} \end{pmatrix} = \begin{pmatrix} 0 & 0 & 0 \\ 0 & 0 & -\vec{y}^{(2)T} \partial_\xi^2 \vec{q} \\ 0 & \vec{y}^{(2)T} \partial_\xi^2 \vec{q} & 0 \end{pmatrix}$$

and from (5.14), together with the co-ordinate normalization (1.10),

$$k_\xi = \begin{pmatrix} 0 & -1 & 0 \\ 1 & 0 & 0 \\ 0 & 0 & 0 \end{pmatrix} .$$

Then

$$ad(k_\xi)^{-1} (A_\xi)_p = \begin{pmatrix} 0 & 0 & -\vec{y}^{(2)T} \partial_\xi^2 \vec{q} \\ 0 & 0 & 0 \\ \vec{y}^{(2)T} \partial_\xi^2 \vec{q} & 0 & 0 \end{pmatrix}$$

and

$$\frac{1}{2} \left[ad(k_\xi)^{-1} (A_\xi)_p , (A_\xi)_p \right]_{\mathcal{A}} = \frac{1}{2} \begin{pmatrix} 0 & -(\vec{y}^{(2)T} \partial_\xi^2 \vec{q})^2 & 0 \\ (\vec{y}^{(2)T} \partial_\xi^2 \vec{q})^2 & 0 & 0 \\ 0 & 0 & 0 \end{pmatrix} .$$

For the right-hand side of (6.11) we get from (5.16) and (2.10)

$$(k_\eta)_{\mathcal{A}} = (g^T D_\eta g)_{\mathcal{A}} = \left(\begin{pmatrix} \vec{q}^T \\ \partial_\xi \vec{q}^T \\ \vec{y}^{(2)T} \end{pmatrix} (\partial_\eta \vec{q}, D_\eta Y) \right)_{\mathcal{A}} = \begin{pmatrix} 0 & -\partial_\xi \vec{q}^T \cdot \partial_\eta \vec{q} & 0 \\ \partial_\xi \vec{q}^T \cdot \partial_\eta \vec{q} & 0 & 0 \\ 0 & 0 & 0 \end{pmatrix} .$$

Expanding $\partial_\xi^2 \vec{q}$ in the basis (5.16) of $\mathbb{R}^3$, we find

$$(\vec{y}^{(1)T} \partial_\xi^2 \vec{q})^2 = (\partial_\xi^2 \vec{q})^2 - 1 ,$$

and (6.11) gives the conservation law (1.12.1)

$$\partial_\eta \frac{1}{2}(\partial_\xi^2 \vec{q})^2 = \partial_\xi(\partial_\xi \vec{q} \cdot \partial_\eta \vec{q}) \quad .$$

The corresponding, much more involved, calculation for the general complex Grassmannian is explained in detail in Ref. 3).

Already the first higher conservation law (6.11) displays the characteristic feature that the currents are rational functions rather than polynomial ones. Due to the $\mathrm{ad}(k_\xi)^{-1}$, which for the nth current will occur in the nth power, there are denominators which can become zero. As the S^2 example shows, these denominators can be eliminated for the rank one spaces by the co-ordinate normalization (1.10). However, for spaces of higher rank, there will necessarily remain rational singularities in the currents.

ACKNOWLEDGMENTS

It is a pleasure to thank M. Forger for his continuous and fruitful collaboration in producing the reported results.

REFERENCES

1) H. Eichenherr and M. Forger, Nucl. Phys. B155 (1979) 381.

2) H. Eichenherr and M. Forger, Nucl. Phys. B164 (1980) 528.

3) H. Eichenherr and M. Forger, "Higher Local Conservation Laws for Non-Linear σ Models on Symmetric Spaces", Freiburg preprint THEP 81/5 (April 1981).

4) M. Forger, "Differential Geometric Methods in Non-Linear σ Models and Gauge Theories", Ph.D. Thesis, Berlin.

5) D. Maison, "Some Facts about Classical Non-Linear σ Models", Munich preprint MPI-PAE/PTh 52/79 (Nov. 1979).

6) H. Eichenherr, in:proceedings of the International Summer Institute on Theoretical Physics, Bad Honnef, Sept. 1980 (Plenum Press,to appear).

7) K. Pohlmeyer, Comm. Math. Phys. 46 (1976) 207.

8) M.J. Ablowitz, D.J. Kaup, A.C. Newell and H. Segur, Stud. Appl. Math. LIII (1974) 249.

9) M. Lüscher and K. Pohlmeyer, Nucl. Phys. B137 (1978) 46.

10) V.E. Zakharov, L.A. Takhtajan, Theor. Math. Phys. 38 (1979) 17.

11) L.D. Faddeev and L.A. Takhtajan, Theor. Math. Phys. 21 (1974) 1046.

12) S. Helgason, Differential Geometry, Lie Groups and Symmetric Spaces (Academic Press, New York 1978).

13) V.E. Zakharov, A.V. Michailov, Sov. Phys. JETP 47 (1978) 1017.

14) K.-H. Rehren and K. Pohlmeyer, J. Math. Phys. 20 (1979) 2628.

15) H. Eichenherr and J. Honerkamp, J. Math. Phys. 22 (1981) 374.

16) M. Caselle, R. Megna and S. Scinto, "Generalizations of the sine-Gordon Equation with $SU(p+q)/S[U(p)\times U(q)]$ Structure", Torino preprint IFTT 402 (Jan. 1981).

17) H. Eichenherr, Phys. Lett. 90B (1980) 121.

M U L T I M O N O P O L E S

E. Corrigan

Department of Mathematics,
University of Durham, U.K.

C o n t e n t s

1. Introduction

The topic of this meeting is integrable quantum field theories, and that really means theories in two space-time dimensions where most of the work has been done. Investigation of field theory in four dimensions is still in its infancy compared with the two dimensional situation and, for that reason, the subject of this talk may be inappropriate.

An interesting property of many two dimensional field theories is the existence of soliton solutions to their equations of motion. That is, particle like field configurations which occur in the Sine-Gordon theory or the Korteweg de Vries equation for example[1]. In four dimensions examples of such structures are not easy to find and their properties have not been explored in much detail. On the other hand, their existence may indicate integrability (in some sense) so their investigation is probably worthwhile.

't Hooft and Polyakov[2] discovered that spontaneously broken gauge theories have finite energy particle-like solutions to their classical equations of motion which can be interpreted naturally as magnetic monopoles. Since then considerable work has been done to generalise these solutions to discover if they really correspond to four dimensional 'solitons'. Until recently, most of the work has involved generalisations of spherically symmetric monopoles to gauge groups other than SU(2). On the one hand, it has proved possible to develop a catalogue of the kinds of magnetic charges that may arise and their properties[3], whilst on the other, exact solutions have been discovered in a rather special situation - the Bogomolny-Prasad-Sommerfield (BPS) limit[12,4]. In this limit the term in the gauge theory lagrangian corresponding to the Higgs' potential is deleted but the appropriate spatially asymptotic boundary condition on the Higgs' field is maintained by hand.

Studying the spherically symmetric solution is in any case interesting because the monopole solution turns out to be constructed from solutions to certain finite dimensional Toda lattices[5]. The latter are themselves integrable systems, widely studied in their own right. They do, however, belong to two space-time dimensions and occur in the gauge theory because the im position of spherical symmetry reduces the number of independent variables in the problem (to one).

An apparently harder problem is to construct finite energy solutions in the spontaneously broken SU(2) theory corresponding to multiparticle states. One feels that such solution should exist with the 'particles' moving around under the influence of effective forces determined consistently by the classical equations of motion for the fields themselves. Such solutions would be time dependent and, very likely, too complicated to describe in much detail.

However, the BPS limit has some very special properties. Manton[6] pointed out that monopoles of like magnetic charge, which one would expect to repel each other according to Coulomb's law (at least if they were widely separated), do not in fact experience a force at all when they are far apart. The long range attractive Higgs' force cancels the long range repulsive coulomb force. Perhaps more astonishing is the existence theorem of Taubes[7] showing that there exist multiply charged (N units of magnetic charge) <u>static</u> solutions in the BPS limit whose total energy is precisely N times the energy of a single (spherically symmetric) monopole solution. Apparently, under these circumstances like charge monopoles are able to rest in equilibrium; albeit a precarious equilibrium destroyed by switching on the Higgs' potential.

Recently, Forgacs, Horvath and Palla[8] and Ward[9] have pioneered the construction of monopole solutions for $N \geqslant 2$ in two very different ways. The first set of authors noted that if we look for axially symmetric monopoles we shall be lead to study the Ernst equation[10].

This is by itself remarkable since the Ernst equation is another two-dimensional field equation which has received a great deal of attention for other reasons. In particular, it has several Bäcklund transformations associated with it[11] which can be used to generate monopoles from the trivial vacuum solution. This is a phenomenon, similar to that happening for the Sine-Gordon solitons which are also generated by Bäcklund transformations[1], which I find very exciting and would like to pursue further. Ward, however, took a quite different approach preferring to make use of some formalism (described below, section 2) which had been invented for the study of instantons. Ward's approach, as will be described below, does not require any specific symmetry to be imposed and, it appears that all the multimonopole solutions may be described naturally within it. For this reason we shall not discuss the work of ref (8) further in these lectures. An understanding of the relationship between the two approaches would be most enlightening.

The situation we are led to consider is thus rather special. Let us denote the SU(2) gauge fields and adjoint Higgs' field by A_μ, Φ respectively, each represented by a 2 × 2 traceless antihermition matrix, a linear combination of the SU(2) generators $\frac{1}{2} i \underline{\sigma}$. Then, the non-abelian 'magnetic' and 'electric' fields are given by

$$H_i = \varepsilon_{ijk} (\partial_j A_k + A_j A_k) = \tfrac{1}{2} \varepsilon_{ijk} F_{jk}$$

$$F_{oi} = E_i = \partial_o A_i - \partial_i A_o + [A_o, A_i], \qquad (1.1)$$

respectively. The theory is described by the lagrangian[3]

$$\mathscr{L} = \tfrac{1}{2} \operatorname{tr} (F^+_{\mu\nu} F_{\mu\nu}) + \operatorname{tr}(D_\mu \Phi)^+ D_\mu \Phi - V(\Phi).$$

In the BPS limit $V(\Phi)$ is deleted and replaced by the asymptotic condition

$$2\ \mathrm{tr}\ \Phi^{+}\Phi \rightarrow 1 \text{ as } |\underline{x}| \rightarrow \infty . \tag{1.2}$$

Moreover, we are seeking solutions for which $E_i = 0$ and all fields are static. In that case the energy is given by,

$$\begin{aligned} \varepsilon &= \int d^3\underline{x}\ \{\mathrm{tr}\ \underline{H}^{+}\underline{H} + \mathrm{tr}((\underline{D}\Phi)^{+}\ \underline{D}\Phi)\} \\ &= \int d^3\underline{x}\ \{\mathrm{tr}(\underline{H} \mp \underline{D}\Phi)^{+}(\underline{H} \mp \underline{D}\Phi) \pm 2\int d^3x\ \mathrm{tr}(\underline{H}^{+}\cdot\underline{D}\Phi) \\ &\geqslant 4\ \pi|N|, \end{aligned} \tag{1.3}$$

where

$$4\pi N = 2\int d^3\underline{x}\ \mathrm{tr}(\underline{H}^{+}\cdot\underline{D}\Phi) = 2\int_{S\infty} \underline{ds}\cdot \mathrm{tr}(\underline{H}^{+}\Phi) \tag{1.4}$$

is the magnetic charge of the configuration.

When $N = 1$, Prasad and Sommerfield(12) found a solution for which $\varepsilon = 4\pi$ satisfying

$$\underline{H} = \pm\underline{D}\Phi, \tag{1.5}$$

and hence saturating the bound (1.3) for the given N(=1). It's worth giving the detailed value of Φ for this solution it is

$$\Phi = \frac{-i\ \hat{\underline{x}}\cdot\underline{\sigma}}{2}(\coth r - \frac{1}{r}) \qquad r = |\underline{x}| \tag{1.6}$$

a function which is everywhere regular, vanishes at the origin, and, asymptotically, behaves like

$$+2\ \mathrm{tr}\ \Phi^{+}\Phi = (\coth r - \frac{1}{r})^2 \simeq (1 - \frac{2}{r}) \quad \text{as } r \rightarrow \infty . \tag{1.7}$$

Because of (1.4) and (1.5) we see that the coefficient of $\frac{1}{r}$ in eq.(1.7) defines the magnetic charge and we expect for more complicated solutions in a regular gauge,

$$\Phi \simeq -\frac{i\ \underline{n}(N)\cdot\underline{\sigma}}{2}(1 - \frac{N}{r}) \quad \text{as } r \rightarrow \infty. \tag{1.8}$$

In eq.(1.8) the unit vector $\underline{n}(N)$ is an angular function (like $\hat{\underline{x}}$) but mapping the sphere at ∞ to the sphere $\underline{n}^2 = 1$ with homotopy class N[3,13]. Eq.(1.8) does not give much of a clue however, as to how to construct the Higgs' field and gauge fields for a given integer N for all values of $\underline{x}$.

A useful observation is that the Bogomolny equations (1.5) are equivalent to self-dual equations[14]. Introduce a fictitious euclidean variable x_4 (x_0 is the time) on which no field depends. Then, setting $\Phi = A_4$ we have

$$F_{i4} = \partial_i A_4 - \partial_4 A_i + [A_i, A_4] = D_i \Phi$$

and solutions to equations (1.5) may be thought of as solutions to the duality equations

$$F_{\mu\nu} = \pm {}^{*}F_{\mu\nu} = \pm \tfrac{1}{2} \varepsilon_{\mu\nu\rho\sigma} F_{\rho\sigma} \qquad \mu,\nu,\rho,\sigma = 1,2,3,4 \tag{1.9}$$

which happen to be independent of x_4. The observation is useful because quite a lot is known about self (antiself) dual fields and, in particular, it is known how to construct all of them which give rise to a finite action - the instantons and anti-instantons[15]. Thus there is a hope that the techniques learnt for the instantons may be helpful for the multi-monopoles also.

Unfortunately, such a hope does not readily materialise. The construction of Atiyah, Drinfeld, Hitchin and Manin[16] which so elegantly yields all the instantons does not appear to be tailored to fit the monopoles as well. This despite the fact, pointed out by Nahm[17], that the N = 1 spherically symmetric monopole can be described rather neatly by a natural generalisation of the ADHM procedure. Up to the present it has not proved possible to find multi-monopoles this way.

However, the ADHM construction is really a way of constructing certain analytic vector bundles over CP_3, a problem which Ward[18], and Atiyah and

and Ward[19], had shown was equivalent to the problem of finding vector potentials yielding self-dual field strengths. These authors showed that eqns. (1.9) could be solved by a series of ansätze a_ℓ $\ell = 1,2,\ldots$ corresponding naturally to classes of 'patching matrices' necessary to the description of the vector bundles. The ansatz a_1 corresponded to the previously known ansatz used by 't Hooft[20] and others to construct a subclass of the multi-instanton solutions. It had also been used by Manton[21] to describe the one monopole solution in the following way.

Let

$$\Delta_o = e^{ix_4} \frac{\text{sh}r}{r}. \tag{1.10}$$

Then, $\partial^2 \Delta_o = 0$ and the a_1 ansatz says that

$$A_\mu = -\eta_{\mu\nu} \partial_\nu \ln \Delta_o \,. \tag{1.11}$$

Hence,

$$\Phi = A_4 = -\frac{i\underline{\sigma} \cdot \underline{\nabla}}{2} \ln \Delta_o = -\frac{i\underline{\sigma} \cdot \hat{\underline{x}}}{2} (\coth r - \frac{1}{r}). \tag{1.12}$$

We may note also that

$$2 \, \text{tr}\, \Phi^+ \Phi = 1 - \nabla^2 \ln \frac{\text{sh} r}{r} \tag{1.13}$$

and, in this picture, the magnetic change is coming from the property

$$\frac{\text{sh} r}{r} \simeq \frac{e^r}{r} \quad \text{as} \quad r \to \infty \,. \tag{1.14}$$

(The $\eta_{\mu\nu}$ appearing in eq.(1.11) is the tensor introduced by 't Hooft[15], eqn.(2.17) below).

Furthermore, Manton showed that only the spherically symmetric monopole could be obtained via this ansatz.

The set of ansätze never seemed helpful in constructing the instantons but with hindsight the seeds of a solution for monopoles exist already in eqns. (1.13) and (1.14). Recently, Ward[9] pointed out how to construct an axisymmetric solution for $N = 2$ and his method was generalised by Prasad and Rossi[22] to certain axisymmetric solutions for any integer N. These solutions made essential use of the ansätze a_ℓ, $\ell > 1$, and, rather remarkably, it was found that the ansatz a_N is required for a monopole solution of total magnetic charge N. It seems that the tower of ansätze - a confusion for the instantons - is actually tailored for monopoles!

However, the axisymmetric solutions do not exhaust the possibilities. In fact they form a very small subset of all solutions. E. Weinberg[23] pointed out that multimonopole solutions to eqns. (1.5) with total magnitude charge N will depend upon $4N - 1$ free parameters in general (for SU(2)). The axisymmetric solutions depend on just five parameters regardless of what value N takes. Nevertheless, it will be shown in section (5) that the axisymmetric solutions are special cases of a general $4N - 1$ parameter set constructed systematically within the ansatz a_N.

In the next section those elements of the Atiyah-Ward ansatz that we need will be described in some detail, and in subsequent sections we shall discuss the multi-monopole solutions themselves. The discussion will not be very rigorous. In particular it will not be possible to demonstrate that the $4N - 1$ parameter set of solutions exposed is complete, or that there are no more constraints on the parameters to ensure non-singularity of the Higgs' field, Φ, and the 'magnetic' gauge field, $\underline{H}$.

2. Atiyah-Ward-Yang formalism

Both Ward[18] and Yang[24] pointed out the importance of complexifying the four Euclidean coordinates x_μ, $\mu = 1,2,3,4$. The crucial observation was that null planes in the complex space (metric $\delta_{\mu\nu}$) were either dual or anti-dual. Having said this they realised that restricted to an anti-dual null plane, a self-dual gauge field is trivial. Ward pointed out that the set of all anti-dual planes could be thought of as a three dimensional complex projective space CP_3 (minus a line) and used this information together with the triviality of the self-dual gauge field on each plane to 'encode' the vector potential into the structure of certain analytic two dimensional vector bundles. Yang's point of view was less geometrical but we shall return to it later. It complements Ward's rather nicely.

It is convenient to write the spatial coordinates in a 2×2 matrix form:

$$x = x_4 - i\underline{x} \cdot \underline{\sigma} = \begin{pmatrix} y & z \\ -\bar{z} & \bar{y} \end{pmatrix} \tag{2.1}$$

with

$$\begin{aligned} iy &= x_3 + ix_4, \quad -i\bar{y} = x_3 - ix_4, \\ iz &= x_1 + ix_2, \quad -i\bar{z} = x_1 - ix_2 . \end{aligned} \tag{2.2}$$

Then the anti-dual null planes are described by pairs of complex two vectors (π,ω) satisfying

$$x\pi = \omega, \tag{2.3}$$

and clearly the same plane is described by $(\lambda\pi,\lambda\omega)$ for any non-zero complex number λ. The vector bundle is defined by its patching function $g(\omega,\pi)$ which is homogeneous:

$$g(\omega,\pi) = g(\lambda\omega,\lambda\pi), \tag{2.4}$$

and satisfies

$$\det g = 1. \tag{2.5}$$

By virtue of eqns.(2.3) and (2.4), the matrix function g can be thought of as a function of μ, ν and ζ, given by

$$\zeta = \frac{\pi_1}{\pi_2}, \qquad \mu = (x_1 + ix_2)\zeta - x_3 + ix_4 = \left(i\frac{\omega_2}{\pi_2} \right) \tag{2.6}$$

and

$$\nu = (x_1 - ix_2)\frac{1}{\zeta} + x_3 + ix_4 = \left(i\frac{\omega_1}{\pi_1} \right).$$

We notice immediately that for fixed ζ,

$$\partial^2 g = 0 \tag{2.7}$$

automatically.

We require that for fixed x_μ, g should be regular in some annulus excluding the points $\zeta = 0, \infty$ in the complex ζ plane. We also require that it be possible to split g multiplicatively into two factors h, k^{-1},

$$g = hk^{-1}, \tag{2.8}$$

where h is regular in a region excluding $\zeta = 0$ and k is regular in a region excluding the point $\zeta = \infty$. In that case the vector potential is given by

$$\begin{aligned} A_{i1} - \zeta A_{i2} &= h^{-1}\left(\frac{\partial}{\partial x_{i1}} - \zeta \frac{\partial}{\partial x_{i2}} \right) h \\ &\equiv k^{-1}\left(\frac{\partial}{\partial x_{i1}} - \zeta \frac{\partial}{\partial x_{i2}} \right) k, \end{aligned} \qquad i = 1,2 \tag{2.9}$$

$(A_{ij}dx_{ij} = (A_\mu dx_\mu))$, the equality of the two pieces following directly from the fact,

$$\left(\frac{\partial}{\partial x_{i1}} - \zeta \frac{\partial}{\partial x_{i2}} \right) g \equiv 0, \quad \text{for} \quad i = 1,2. \tag{2.10}$$

Two such patching functions g, g' will give rise to gauge equivalent vector potentials provided they are related in the following way:

$$g = Ag'a \tag{2.11}$$

where A, a are both functions of μ,ν,ζ alone and, A is regular in a region excluding $\zeta = 0$ whilst a is regular in a region excluding $\zeta = \infty$.

Atiyah and Ward[19] argue that any suitable g will be gauge equivalent in the above sense to a matrix of the form

$$\begin{pmatrix} \zeta^{\ell} & \rho(\mu,\nu,\zeta) \\ 0 & \zeta^{-\ell} \end{pmatrix} \tag{2.12}$$

for some integer $\ell = 1,2,3,\ldots$, where ρ is, as indicated, a function of μ, ν and ζ.

The function ρ will have a Laurent expansion in ζ with coefficients that are functions of x_μ :

$$\rho \;=\; \sum_{-\infty}^{\infty} \zeta^{-r}\,\Delta_r \tag{2.13}$$

and, as a consequence of eq. (2.10)

$$\frac{\partial}{\partial z}\Delta_r \;=\; \frac{\partial}{\partial \bar{y}}\Delta_{r+1}, \quad \frac{\partial}{\partial y}\Delta_r \;=\; -\frac{\partial}{\partial \bar{z}}\Delta_{r+1}\,, \tag{2.14}$$

which in turn implies

$$\partial^2\Delta_r \;=\; 0, \tag{2.15}$$

for every value of r.

It was shown in ref. (25) that eqns. (2.13), (2.8) and (2.9) lead to the following expression for the vector potential (in a special gauge). Let

$$A \;=\; -\frac{i}{2f}\begin{pmatrix} \eta^{3}_{\alpha\beta}\,\partial_\beta f & \eta^{1-i2}_{\alpha\beta}\,\partial_\beta e \\ \eta^{1+i2}_{\alpha\beta}\,\partial_\beta g & -\eta^{3}_{\alpha\beta}\,\partial_\beta f \end{pmatrix} \tag{2.16}$$

where the 't Hooft tensors $\eta^a_{\mu\nu}$ are defined by

$$\eta^a_{\mu\nu} = \varepsilon_{4a\mu\nu} + \delta_{a\mu}\,\delta_{\nu4} - \delta_{a\nu}\,\delta_{\mu4}, \quad \eta^{1\pm i2} = \eta^1 \pm \eta^2, \tag{2.17}$$

Then the functions e,f,g are defined in terms of the coefficients in the

Laurent expansion of ρ, eqn. (2.13), by the recipe:

$$(\text{if } \ell \geq 2) \quad \begin{pmatrix} e & f \\ f & g \end{pmatrix} = \text{outside corner elements of} \begin{pmatrix} \Delta_{-\ell+1} & \cdots & \Delta_2 & \Delta_{-1} & \Delta_0 \\ \cdot & & \Delta_{-1} & \Delta_0 & \Delta_1 \\ \cdot & & \cdot & \Delta_1 & \Delta_2 \\ \cdot & \cdot & & \Delta_2 & \cdot \\ \Delta_{-1} & \Delta_0 & & & \cdot \\ \Delta_0 & \Delta_1 & \cdots & & \Delta_{\ell-1} \end{pmatrix}^{-1} \tag{2.18}$$

If $\ell = 1$ then $e = f = g = 1/\Delta_0$ and eqn.(2.16) reduces to the familiar ansatz of eqn.(1.11).

So far so good but there are a couple of problems. One is that because we have complexified everything we are in danger of producing a self-dual SL(2,C) gauge potential, via eqns.(2.16)-(2.18), instead of what we actually want which is a potential for SU(2). Secondly, there is nothing discernible so far in the construction which will tell us which choice of ρ will lead to either: finite action (instantons), or finite energy (monopoles). The former question has never been resolved but the latter turns out to be more susceptible to analysis.

In order to make sure that when x_μ is real the gauge potential belongs to the SU(2) Lie algebra we note first that when x_μ is real

$$\bar{y} = y^*$$

$$\bar{z} = z^*$$

where * denotes complex conjugation (the bar does not) and so

$$A_{11}^{+} = -A_{22}, \quad A_{12}^{+} = A_{21} . \tag{2.19}$$

Hence, if the vector potential corresponds to SU(2) there must be a condition following from the form of the defining equation (2.9). Explicitly it is[19]

$$\left[h(-\tfrac{1}{\zeta^*}) \right]^{+} = k^{-1}(\zeta), \tag{2.20}$$

and, in terms of the patching matrix, we have

$$[g(-\frac{1}{\zeta^*})]^+ = g(\zeta) \qquad (2.21)$$

following directly from eqn.(2.8).

Eqn.(2.21) looks straightforward but in fact it is not. The reason is that the form of g which is convenient for computation is the upper triangular form of eqn.(2.12) - which always violates eqn.(2.21). Actually, it is enough that the upper triangular form be gauge equivalent, (in the sense of eqn.(2.11)), to a matrix which satisfies what we may call the reality condition, (2.21). We shall see that this condition is very strong and not at all easy to satisfy. The mismatch between the upper triangular patching matrix and one that satisfies the reality condition is a nuisance in practice because the result of the computation starting with a given ρ will generally yield vector potentials in a complex gauge.

To obtain some idea of how to choose a suitable ρ it is useful to go back and consider the alternative development initiated by Yang.

Yang's starting point[24,26] was much the same as Ward's but he proceeded more analytically than geometrically, pointing out that in terms of the complex coordinates $y,\bar{y},z,\bar{z}$ the self duality equations (1.9) read:

$$F_{yz} = 0, \quad F_{\bar{y}\bar{z}} = 0, \quad F_{y\bar{y}} + F_{z\bar{z}} = 0. \qquad (2.22)$$

In other words, the components A_y $(=A_{11})$ and A_z $(=A_{12})$ must be given by

$$A_y = D^{-1}\partial_y D, \quad A_z = D^{-1}\partial_z D \qquad (2.8)$$

and, similarly,

$$A_{\bar{y}} = \bar{D}^{-1}\partial_{\bar{y}}\bar{D}, \quad A_{\bar{z}} = \bar{D}^{-1}\partial_{\bar{z}}\bar{D}, \qquad (2.24)$$

the matrices D and $\bar{D}$ being arbitrary SL(2,C) matrices. We remark that the combination $J = D\bar{D}^{-1}$ is gauge invariant and that it may be parameterised in the following way

$$J = \frac{1}{\phi}\begin{pmatrix} 1 & \gamma \\ -\varepsilon & \phi^2 - \varepsilon\gamma \end{pmatrix}, \qquad (2.25)$$

where ε, ϕ and γ are three complex functions of the coordinates.

The matrix J conveniently splits into two factors

$$J = \frac{1}{\sqrt{\phi}}\begin{pmatrix} 1 & 0 \\ -\varepsilon & \phi \end{pmatrix} \frac{1}{\sqrt{\phi}}\begin{pmatrix} 1 & \gamma \\ 0 & \phi \end{pmatrix} \qquad (2.26)$$

which are to be identified with D and $\bar{D}^{-1}$ respectively. (In so doing we have effectively selected a gauge.) The first two equations (2.22) are satisfied identically but the third requires ε, ϕ, γ to satisfy:

$$\partial^2 \ln\phi = \frac{4}{\phi^2}(\varepsilon_y\gamma_{\bar{y}} + \varepsilon_z\gamma_{\bar{z}}) \qquad (2.27)(i)$$

$$(\varepsilon_y/\phi^2)_{\bar{y}} + (\varepsilon_z/\phi^2)_{\bar{z}} = 0 \qquad (ii)$$

$$(\gamma_{\bar{y}}/\phi^2)_{y} + (\gamma_{\bar{z}}/\phi^2)_{z} = 0, \qquad (iii)$$

a rather complicated set of equations with a number of remarkable properties. Many authors[25,26] have discussed these equations and elucidated some of their properties, we shall be interested in just two of them[25].

(I) If ε, ϕ, γ are a solution of eqns.(2.27) then so also are e, f, g, related to ε, ϕ, γ by:

$$\begin{pmatrix} e & f \\ f & g \end{pmatrix} = \begin{pmatrix} \varepsilon & \phi \\ \phi & \gamma \end{pmatrix}^{-1}. \qquad (2.28)$$

Furthermore, the gauge potentials defined via eqns.(2.26), (2.23) and (2.24) in terms of ε, ϕ, γ or e, f, g are gauge transformations of each other with the gauge matrix k given by

$$k = \begin{pmatrix} \gamma & \phi \\ \phi & \varepsilon \end{pmatrix} \frac{1}{\sqrt{\phi^2 - \varepsilon\gamma}}. \qquad (2.29)$$

(II) If ε, ϕ, γ are a solution of eqns.(2.27) then so also are e, f, g, related to ε, ϕ, γ by the set of differential equations:

$$\phi = 1/f$$

$$\varepsilon_z = -\frac{1}{f^2} g_{\bar{y}} \qquad \varepsilon_y = \frac{1}{f^2} g_{\bar{z}} \tag{2.30}$$

$$\gamma_{\bar{z}} = \frac{1}{f^2} e_y \qquad \gamma_{\bar{y}} = -\frac{1}{f^2} e_z \ .$$

One interesting consequence of eqns.(2.30) is the following. If we calculate the vector potential, via equations (2.23) and (2.24), in terms of ε, ϕ and γ then use eqn.(2.30) to re-express the result in terms of e, f and g the answer comes out to be precisely the expression for the vector potential given in eqn.(2.16). The two choices of gauge are the same.

Another consequence of properties (I) and (II) taken together is an iterative proof of the formula for e, f, g given in eqn.(2.18). It is easy to convince oneself of that by starting with $e_1 = f_1 = g_1 = 1/\Delta_0$. Using eqn.(2.30) to define two new functions Δ_1 and Δ_{-1} (satisfying eqns.(2.14) with respect to Δ_0) followed by eqn.(2.28) we obtain

$$\begin{pmatrix} e_2 & f_2 \\ f_2 & g_2 \end{pmatrix} = \begin{pmatrix} \Delta_{-1} & \Delta_0 \\ \Delta_0 & \Delta_1 \end{pmatrix}^{-1} , \tag{2.31}$$

as desired. The details of the inductive step leading to a complete proof of eqn.(2.18) can be found in ref.(25).

Some time ago, Lohe[(27)] tried to use the 'Bäcklund' transformation (2.30) to generate multimonopole solutions starting from a guess for Δ_0. Unfortunately that was not successful, the main reason being that although for a given function ρ (and hence a set of Δ_r's) one can obtain vector potentials within each ansatz, they will in general be singular and lead to singular field strengths, $F_{\mu\nu}$. Rather, the choice of ρ (for a given value of ℓ) leading to regular solutions is a delicate matter, only partially understood as we shall see.

In the next section we shall take some steps towards finding a generalisation of the simple solution described in eqns.(1.10)-(1.14). The hint, already there in equations (1.13) and (1.14), is that we should seek a sort of multiplicative superposition principle for the argument of the logarithm which, because of the logarithm, becomes additive in equation (1.13). At least for large r we should like to find

$$2\,\mathrm{tr}\,\Phi^+\Phi \;\simeq\; 1 - \nabla^2 \ln \prod_{i=1}^{N} \Delta_i, \quad \text{as } r \to \infty , \tag{2.32}$$

where each Δ_i has an asymptotic behaviour no more (or less) than exponential

$$\Delta_i \sim e^r f_i(r,\theta,\phi), \quad i = 1,\ldots,N.$$

In that case, the total magnetic charge of the solution will come out to be N. Remarkably, there is such a formula, due to Prasad, which provides us with the generalisation of eqn.(1.13) that we need.

3. Prasad's formula for $\mathrm{tr}\,\Phi^+\Phi$ (28)

From eqn.(2.16) we can calculate the Higgs' field to be

$$\Phi \equiv A_4 = \frac{i}{2f}\begin{pmatrix} \partial_3 f & -2i\,\partial_z e \\ 2i\,\partial_{\bar z} g & -\partial_3 f \end{pmatrix}. \tag{3.1}$$

So we also have:

$$2\,\mathrm{tr}\,\Phi^+\Phi = \frac{(\partial_3 f)^2 + 4\,\partial_z e\,\partial_{\bar z} g\,)}{(f)^2}. \tag{3.2}$$

This does not look very useful but, let us for a moment specialize to the case of $\ell = 2$ for which

$$\begin{pmatrix} e_2 & f_2 \\ f_2 & g_2 \end{pmatrix} = \begin{pmatrix} \Delta_{-1} & \Delta_0 \\ \Delta_0 & \Delta_1 \end{pmatrix}^{-1}, \tag{3.3}$$

remarking that

$$\begin{aligned} 2\,\mathrm{tr}(\Phi_2^+\Phi_2) &= \frac{(\partial_3 f_2)^2 - 4\partial_y e_2 \partial_{\bar y} g_2}{(f_2)^2} + \frac{4(\partial_z e_2 \partial_{\bar z} g_2 + \partial_y e_2 \partial_{\bar y} g_2)}{(f_2)^2} \\ &= \frac{(\partial_3 f_2)^2 - 4\partial_y e_2 \partial_{\bar y} g_2}{(f_2)^2} + \partial^2 \ln f_2 \,. \end{aligned} \tag{3.4}$$

In the second step we have used eqn.(2.24(i)). Now, compare eqn.(3.4) with the expression for the Higgs' field in the $\ell = 1$ ansatz:

$$2\,\mathrm{tr}(\Phi_1^+\Phi_1) = \frac{(\partial_3 f_1)^2 + 4\partial_z e_1 \partial_{\bar z} g_1}{(f_1)^2} = 1 - \nabla^2 \ln \Delta_0 .$$

$$\left(\nabla^2 = \sum_1^3 \partial_i^2\right) \tag{3.5}$$

where, in the last equality we have assumed already that Δ_0 has the x_4 dependence given by eqn.(1.10) and that $e_1 = f_1 = g_1 = 1/\Delta_0$. Using eqns.(2.14) we have:

$$2\,\mathrm{tr}(\Phi_1^+\Phi_1) = \frac{(\partial_3\Delta_0)^2 - 4\partial_y \Delta_{-1} \partial_{\bar y} \Delta_1}{\Delta_0^2} \tag{3.6}$$

the expression we would have got using Yang's variables via eqns.(2.23) and (2.24). Finally, we remember our previous assertion that making the replacement defined by eqn.(3.3) is actually a gauge transformation in Yang's description, and will have no effect whatsoever on the value of $\mathrm{tr}\,\Phi_1^+\Phi_1$, provided the change of gauge is independent of x_4. That this is indeed the case is guaranteed by the assumption that

$$\begin{pmatrix} \Delta_{-1} & \Delta_0 \\ \Delta_0 & \Delta_1 \end{pmatrix} = e^{ix_4} \begin{pmatrix} \tilde\Delta_{-1} & \tilde\Delta_0 \\ \tilde\Delta_0 & \tilde\Delta_1 \end{pmatrix}, \tag{3.7}$$

since all dependences on x_4 in the gauge matrix k, (eqn.(2.29)), conveniently cancel away.

Combining this remark with eqns.(3.6), (3.5) and (3.4) we find

$$\begin{aligned} 2\,\mathrm{tr}(\Phi_2^+ \Phi_2) &= 2\,\mathrm{tr}\,\Phi_1^+ \Phi_1 + \partial^2 \ln f_2 \\ &= 1 - \nabla^2 \ln \Delta_0 + \nabla^2 \ln \left(\frac{\Delta_0}{\Delta_0^2 - \Delta_1 \Delta_{-1}} \right) \\ &= 1 - \nabla^2 \ln (\Delta_0^2 - \Delta_1 \Delta_{-1}) . \end{aligned} \qquad (3.8)$$

Indeed, using a combination of the transformations (i) and (ii) of section (2) Prasad was able to deduce

$$2\,\mathrm{tr}\,\Phi_\ell^+ \Phi_\ell = 2\,\mathrm{tr}\,\Phi_{\ell-1}^+ \Phi_{\ell-1} + \partial^2 \ln f_\ell ,$$

and hence that

$$2\,\mathrm{tr}(\Phi_\ell^+ \Phi_\ell) = 1 - \nabla^2 \ln \det D_\ell, \qquad (3.9)$$

where D_ℓ is the banded matrix of Δ's appearing in the formula (2.18),

$$(D_\ell)_{ij} = \Delta_{-i+j+1-\ell} \quad \text{for} \quad i,j = 1,2,\ldots,\ell . \qquad (3.10)$$

(Notice that the formula (3.9) bears a striking resemblance to multisoliton formulae in classical soliton theory[1].)

It must be stressed, however, that the dependence on x_4 has to be special:

$$\Delta_r = e^{ix_4} \tilde{\Delta}_r, \quad \forall r, \qquad (3.11)$$

where $\tilde{\Delta}_r$ depends only on x_1, x_2 and x_3. Any other dependence on x_4 appears to ruin the elegant formula (3.9).

Notice that the functions $\tilde{\Delta}_r$ must satisfy the equation

$$\nabla^2 \tilde{\Delta}_r = \tilde{\Delta}_r , \quad \forall r, \qquad (3.12)$$

which actually implies (unless the angular behaviour is peculiar) that for large r

$$\tilde{\Delta}_r \simeq \frac{e^r}{r} f(\theta,\phi). \qquad (3.13)$$

This asymptotic behaviour is exactly what we need for monopole solutions of total charge N, and the formula (3.9) strongly suggests that only for $\ell = N$ can we hope to find such a solution. As mentioned before, it appears the Atiyah-Ward ansatze are put together in such a way as to incorporate the monopole boundary conditions quite naturally, rather than those for the instantons.

4. The choice of patching function

The special x_4 dependence of the Δ's has an immediate consequence for the form of the function ρ which appears in the patching function in its upper triangular form, eqn.(2.12). The point is that although the condition on the Δ's (eqn.3.11) appears to imply

$$\rho = e^{ix_4}\tilde{\rho} \quad ,$$

it is actually much stronger. The reason is that the general arguments of Ward, indicated in section (2) imply that ρ depends on the coordinates x_μ only in certain combinations with ζ, namely the μ and ν variables of eqn.(2.6). In other words we must in fact have

$$\rho = \exp\left(\frac{\mu+\nu}{2}\right)\tilde{\rho}\left(\frac{\mu-\nu}{2}, \zeta\right), \tag{4.1}$$

where $\tilde{\rho}$ depends upon x_1, x_2 and x_3 in the combination

$$2\gamma \equiv \mu - \nu = (x_1 + ix_2)\zeta - 2x_3 - (x_1 - ix_2)\frac{1}{\zeta} \quad , \tag{4.2}$$

and upon ζ.

Hence the patching matrix must take the form

$$g = \begin{pmatrix} \zeta^{\ell} & \exp\left(\frac{\mu+\nu}{2}\right)\tilde{\rho}(\gamma,\zeta) \\ 0 & \zeta^{-\ell} \end{pmatrix} \tag{4.3}$$

which is easily seen to be gauge equivalent (in the sense of eqn.(2.11) to

$$g' = \begin{pmatrix} \zeta^{\ell} e^{+\gamma} & \tilde{\rho} \\ 0 & \zeta^{-\ell} e^{-\gamma} \end{pmatrix}. \tag{4.4}$$

[The equivalence is achieved by choosing $A = \begin{pmatrix} e^{\nu/2} & 0 \\ 0 & e^{-\nu/2} \end{pmatrix}$, regular except at $\zeta = 0$, and $a = \begin{pmatrix} e^{-\mu/2} & 0 \\ 0 & e^{\mu/2} \end{pmatrix}$, regular except at $\zeta = \infty$.]

We now have to seek a systematic way of choosing $\tilde{\rho}$ so that the patching matrix is gauge equivalent to a matrix satisfying the reality condition (2.21). We also have to ensure that the Laurent coefficients of ρ are non-singular and that the determinant of the banded matrix, D_ℓ, is non-zero. This last condition is necessary for the eventual non-singularity of the Higgs' and gauge fields and will be the most difficult to implement. Before doing this, however, we can look at some special cases.

4.1 Ward(9) pointed out that the choice

$$\tilde{\rho} = \frac{\text{sh}\gamma}{\gamma}, \quad \ell = 1 \tag{4.5}$$

leads to the Prasad-Sommerfield monopole. To see this note that

$$\begin{pmatrix} \zeta e^{\gamma} & \frac{\text{sh}\gamma}{\gamma} \\ 0 & \zeta^{-1} e^{-\gamma} \end{pmatrix} = \begin{pmatrix} \frac{\text{sh}\gamma}{\gamma} & \zeta e^{-\gamma} \\ -\frac{1}{\zeta} e^{-\gamma} & \gamma e^{-\gamma} \end{pmatrix} \begin{pmatrix} \gamma\zeta & 1 \\ +1 & 0 \end{pmatrix} \tag{4.6}$$

and

$$\left[\gamma\left(-\frac{1}{\zeta^*}\right)\right]^* = \gamma(\zeta), \quad (\text{if } x_1, x_2, x_3 \text{ are real}) \tag{4.7}$$

so that the patching matrix is equivalent to one satisfying the reality condition. Also,

$$\Delta_0 = \frac{1}{2\pi i} \oint_C \frac{d\zeta}{\zeta} \exp\left(\frac{\mu+\nu}{2}\right) \frac{\text{sh}\gamma}{\gamma} = e^{ix_4} \frac{\text{shr}}{r} \tag{4.8}$$

where the contour C is any contour enclosing the point $\zeta = 0$. Eqn.(4.8) is precisely what we require.

Notice that the contour C is unambiguous because the function $\frac{\text{sh}\gamma}{\gamma}$ is not singular at $\gamma = 0$. In fact, a condition on $\tilde{\rho}$ to ensure non-singularity of the Laurent coefficients of ρ as functions of x_1, x_2 and x_3 is that it should have no singularities in ζ which move with the spatial coordinates $\underline{x}$. If this is so, then the contour does not have to be chosen in a special way and there is no danger of it being 'trapped' by moving singularities in the ζ plane.

4.2 Ward[9] also pointed out that the choice

$$\tilde{\rho} = \frac{\text{ch}\gamma}{\gamma^2 + (\pi^2/4)}, \quad \ell = 2 \tag{4.9}$$

leads to a monopole of charge 2. Again, the patching matrix is equivalent to one which obeys the reality condition by an argument very similar to that of eqn.(4.6),(4.7). Also, the zeros of the denominator of $\tilde{\rho}$ at $\gamma = \pm i\pi/2$ are exactly cancelled by zeros in the numerator. (This would also be the case for a denominator of the form $\gamma^2 + \frac{k^2\pi^2}{4}$ with k any odd integer. However, these are not allowed for the more subtle reason that Δ_0, $\Delta_0^2 - \Delta_1\Delta_{-1}$ are not permitted to vanish.)

A straightforward computation yields:

$$\begin{aligned} \Delta_1 &= -\frac{i}{z}(\Lambda + \partial_3\Lambda)\, e^{ix_4} \\ \Delta_0 &= \frac{2}{z}\partial_{\bar{z}}\Lambda\, e^{ix_4} \\ \Delta_{-1} &= \frac{i}{\bar{z}}(\Lambda - \partial_3\Lambda)\, e^{ix_4} \end{aligned} \tag{4.10}$$

where Λ is given by

$$\Lambda = \text{ch}R + \text{ch}R^* \tag{4.11}$$

with

$$R^2 = \left| \underline{x} + \frac{i\pi}{2} \hat{\underline{z}} \right|^2 = (r^2 - \frac{\pi^2}{4}) + 2i\, x_3 (\frac{\pi}{2}).$$

Notice, Δ_0 is given by

$$\Delta_0 = \frac{\mathrm{sh}R}{R} + \frac{\mathrm{sh}R^*}{R^*} \qquad (4.12)$$

and has the form of two Prasad-Sommerfield solutions displaced from each other by a complex distance $i\pi$ in the x_3 direction. A direct calculation reveals that neither Δ_0 nor $\Delta_1\Delta_{-1} - \Delta_0^2$ vanishes anywhere and that $\Delta_1\Delta_{-1} - \Delta_0^2$ has the correct asymptotic behaviour for total charge 2.

Monopoles are centred at places where the Higgs' field vanishes or, from eqn.(3.9), where

$$1 = \nabla^2 \ln \det D_\ell . \qquad (4.13)$$

In this case, the Higgs' field vanishes only at the origin $r = 0$. Furthermore, the solution can be generalised by translations and rotations to one containing five parameters - which is two less than the general solution for $N = 2$, according to ref (23). Ward(29) has subsequently found a generalisation of eqn.(4.9) containing two extra parameters leading to a solution for $N = 2$ which is not axisymmetric, and for which the Higgs' field vanishes at two points. We shall defer a more detailed discussion of this solution until later.

4.3 Prasad and Rossi(28,22) suggest that the following choices of $\tilde{\rho}$ yield generalisations of the axisymmetric monopoles for $N > 2$. For N even let

$$\ell = N, \quad \tilde{\rho} = \mathrm{ch}\, \frac{\pi}{2}\gamma \Big/ \prod_{m=1}^{N/2} (\gamma^2 + (2m-1)^2) \qquad (4.14)$$

while for N odd let

$$\ell = N, \quad \tilde{\rho} = \mathrm{sh}\, \pi\gamma \Big/ \gamma \prod_{m=1}^{\frac{N-1}{2}} (\gamma^2 + m^2) , \qquad (4.15)$$

then it is clear that the patching matrix is equivalent in either case to one satisfying the reality condition. It is also clear that the zeros in the denominators of either (4.14) or (4.15) are cancelled so that the Δ's are non-singular. It is possible to obtain nice expressions for the Δ's in a straightforward way but it is not at all clear that the solutions so obtained are regular everywhere. There are however indirect arguments which give us some confidence in the non-singularity of the Higgs' and gauge fields obtained from expressions like (4.14) or (4.15). The arguments are given in detail in ref. (22). The axisymmetric solutions §4.2,4.3 have also been obtained independently and in a different way by Forgacs, Horvath and Palla[8,30].

4.4 A general argument for the form of the patching matrix[31]

At the start of section 4 we were led, by considering the x_4 dependence of ρ, to an upper triangular form for the patching matrix,

$$g = \begin{pmatrix} \zeta^{\ell} e^{\gamma} & f(\gamma,\zeta) \\ 0 & \zeta^{-\ell} e^{-\gamma} \end{pmatrix}. \tag{4.16}$$

We have also noted that we should require f to have no singularities in the plane that move with the spatial coordinates x_1, x_2, and x_3.

In order to implement the reality condition we have to find matrices A, a, as in eqn.(2.11), so that

$$\left[g\left(-\frac{1}{\zeta}*\right)\right]^{+} = Ag(\zeta)a. \tag{4.17}$$

In addition, to maintain the x_4 independence, A and a are functions not just of μ and ν but of the combination γ. Perhaps surprisingly eqn.(4.17) is very strong. To see this let

$$A^{-1} = \begin{pmatrix} a & b \\ c & d \end{pmatrix}, \quad a = \begin{pmatrix} s & t \\ u & v \end{pmatrix} \quad ad - bc = 1, \quad sv - ut = 1 \tag{4.18}$$

then in detail we have

$$\begin{pmatrix} a & b \\ c & d \end{pmatrix}\begin{pmatrix} (-)^{\ell}\zeta^{-\ell}e^{\gamma} & 0 \\ [f(-\frac{1}{\zeta^*})]^* & (-)^{\ell}\zeta^{\ell}e^{-\gamma} \end{pmatrix} = \begin{pmatrix} \zeta^{\ell}e^{\gamma} & f(\zeta) \\ 0 & \zeta^{-\ell}e^{-\gamma} \end{pmatrix}\begin{pmatrix} s & t \\ u & v \end{pmatrix} . \tag{4.19}$$

From eqn.(4.19) we can deduce immediately the following.

The functions d and v are related by

$$d(-)^{\ell}\zeta^{\ell} = v\zeta^{-\ell} , \tag{4.20}$$

from which we deduce d and v must each be a polynomial of degree 2ℓ, d as a function of $1/\zeta$ and v as a function of ζ. Having established v to be a polynomial in ζ we remember it is also a function of γ. Without loss of any generality we may write

$$v/\zeta^{\ell} = \prod_{i=1}^{\ell} (\gamma - \gamma_i) \tag{4.21}$$

where the γ_i are functions of ζ but not of x_1, x_2, or x_3. From eqn.(4.19) we also discover

$$\begin{aligned} f(\gamma,\zeta) &= \frac{(-)^{\ell}b\,e^{-\gamma} - te^{\gamma}}{v/\zeta^{\ell}} \\ &= ((-)^{\ell}b\,e^{-\gamma} - te^{\gamma}) \Big/ \prod_{1}^{\ell}(\gamma - \gamma_i) , \end{aligned} \tag{4.22}$$

and

$$\begin{aligned} [f(\gamma(-\tfrac{1}{\zeta^*}), -\tfrac{1}{\zeta^*})]^* &= \frac{ue^{-\gamma} - (-)^{\ell}\,ce^{\gamma}}{d\zeta^{\ell}} \\ &= ((-)^{\ell}ue^{-\gamma} - ce^{\gamma}) \Big/ \prod_{1}^{\ell} (\gamma - \gamma_i) . \end{aligned} \tag{4.23}$$

The fourth equation derivable from eqn.(4.19) is implied by (4.20), (4.21) and (4.22) and the determinant conditions on A and a. We need not consider it further.

The two expressions for the function f, eqns.(4.22) and (4.23), are not automatically compatible but, forcing them to be so yields extra information about b, c, u and t. Because γ satisfies $[\gamma(-\frac{1}{\zeta^*})]^* = \gamma(\zeta)$ we must require

$$\left[\prod_1^{\ell}(\gamma(-\tfrac{1}{\zeta^*}) - \gamma_i(-\tfrac{1}{\zeta^*}))\right]^* = \prod_1^{\ell}(\gamma(\zeta) - \gamma_i(\zeta)) \qquad (4.24)$$

which tells us something about certain combinations of the unknown functions $\gamma_i(\zeta)$. Explicitly, we deduce that

$$\sum_{i_1 < \ldots < i_k = 1}^{\ell} \gamma_{i_1} \cdots \gamma_{i_k} = Q_k(\zeta), \qquad k = 1,\ldots,\ell \qquad (4.25)$$

where $Q_k(\zeta)$ is a polynomial in ζ and $\frac{1}{\zeta}$ of degree k, satisfying

$$(Q_k(-\tfrac{1}{\zeta^*}))^* = Q_k(\zeta). \qquad (4.26)$$

Thus $\sum_1^{\ell} \gamma_i = a_1\zeta + b_1 - a_1^*/\zeta \ , \ b_1 = b_1^*$

$$\sum_{\substack{1 \\ i<j}}^{\ell} \gamma_i\gamma_j = a_2\zeta^2 + b_2\zeta + c_2 - b_2^*/\zeta + a_2^*/\zeta^2 \ , \quad c_2 = c_2^* \qquad (4.27)$$

etc.

Polynomials satisfying eqn. (4.26) of degree k contain 2k + 1 arbitrary real parameters and so the function occurring in the denomination of eqn. (4.23) or (eqn.(4.22) contains

$$\sum_1^{\ell} 2k + 1 = \ell(\ell + 2) \qquad (4.28)$$

real parameters altogether.

There is also a condition on b, c, u and t:

$$(-)^{\ell}[b(-\tfrac{1}{\zeta^*})]^* - (-)^{\ell}u(\zeta) = e^{2\gamma}([t(-\tfrac{1}{\zeta^*})]^* - c(\zeta)). \qquad (4.29)$$

We remark that the unknown pieces of the equation (4.29) have the following property. On the left hand side we have a function which is analytic at the origin and has an expansion in positive powers of ζ. On the right hand side we have (apart from the $e^{2\gamma}$ function) a function which is analytic at $\zeta = \infty$ and has an expansion in positive powers of ζ^{-1}. In addition, each unknown piece can only depend on x_1, x_2 and x_3 via the combination γ. These pieces of information taken together imply that each side of eqn.(4.29) vanishes by itself. Therefore, we must have,

$$\left[b\left(-\frac{1}{\zeta^*}\right)\right]^* = u(\zeta)$$

$$\left[t\left(-\frac{1}{\zeta^*}\right)\right]^* = c(\zeta).$$

However, we still have to determine b and t.

Looking at eqn.(4.22) for arbitrary b and t there is a danger that the function $f(\gamma,\zeta)$ will have poles in the ζ plane whose position depends upon x_1, x_2 and x_3. This is because the values of ζ for which γ becomes equal to one of the γ_i, $i = 1,\ldots,\ell$, clearly depend upon the spatial variables (and will not, by the way, be easy to find in general). To avoid this possibility we must arrange b and t so that the combination

$$(-)^{\ell}\,\frac{b}{t}\,e^{-2\gamma} \tag{4.30}$$

takes the value unity whenever ζ takes one of the special values leading to $\gamma = \gamma_i$, for some i. A minimal way to arrange this cancellation may be concocted as follows.

Let $P(\gamma,\zeta)$ be a polynomial of degree $\ell - 1$ in γ with coefficients which are functions of ζ only, i.e.,

$$P(\gamma,\zeta) = \sum_{k=0}^{\ell-1} a_k(\zeta)\gamma^k .$$

Then, arrange that the combination $P(\gamma,\zeta) - 2\gamma$ goes through special values as γ takes each of the values γ_i. In other words, for ℓ odd we require

$P(\alpha,\zeta) - 2\gamma$ to take values which are even multiples of $i\pi$, whilst for ℓ even we require it to take values which are odd multiples of $i\pi$. This means that the coefficients a_k, $k = 0,\ldots,\ell-1$ are determined by the γ_i for $i = 1,\ldots,\ell$, and, any restrictions on the coefficients a_k will provide implicitly, constraints on the functions γ_i. Clearly, the combination (4.30) will have the correct property provided we set

$$\frac{b}{t} = e^{P(\gamma,\zeta)}, \qquad (4.31)$$

Recalling that b is a function of ζ and γ regular at $\zeta = \infty$, and that t is a function of ζ and γ regular at $\zeta = 0$, we can determine b and t (up to constant factors) provided we can split e^P into two parts - one analytic at $\zeta = 0$ and the other analytic at $\zeta = \infty$, both dependent on ζ and γ only. Writing,

$$a_k(\zeta) = \sum_{r=-\infty}^{\infty} a_k^{(r)} \zeta^{+r} \equiv a_k^{<} + a_k^{o} + a_k^{>},$$

where the superscripts $^{>}$ $^{<}$ or ^{o} indicate the splitting of a_k into its pieces with expansions in ζ, $\frac{1}{\zeta}$ or the constant part respectively, we have

$$P(\gamma,\zeta) = \sum_{k=0}^{\ell-1} \gamma^k (a_k^{<} + a_k^{o} + a_k^{>}) \qquad (4.32)$$

which we want to split into similar pieces $P^{<}$, P^{o} and $P^{>}$. However, γ contains both ζ and $\frac{1}{\zeta}$ and we recognise that the expression (4.32) can only split in the way we want, without dismantling γ, provided

$$a_k^{o}, a_k^{\pm 1}, a_k^{\pm 2}, \ldots, a_k^{\pm(k-1)} = 0 \quad \text{for } 1 \leq k \leq \ell-1. \qquad (4.33)$$

Eqn.(4.33) amounts to a set of constraints on the γ_i and the total number of constraints is given by

$$\sum_{1}^{\ell-1} (2r-1) = (\ell-1)^2 .$$

In other words, the effective number of degrees of freedom in f, if we adopt this minimal procedure, is

$$\ell(\ell+2) - (\ell-1)^2 = 4\ell - 1,$$

precisely the number we want. That this minimal procedure is in fact the correct one may well have to do with the conditions that have to be imposed on ρ to guarantee that determinants of the bounded matrices constructed from its central $2\ell - 1$ Laurent coefficients, eqn.(3.10), do not vanish. The precise nature of these conditions is still obscure and the object of further study.

Returning to eqn.(4.31), and assuming the constraints (4.33) are indeed satisfied, we may write

$$b = \exp(P^< + \tfrac{1}{2}P_o), \qquad t = \exp(-P^> - \tfrac{1}{2}P_o) \, .$$

In which case we define

$$\begin{aligned} f(\gamma,\zeta) &= \left(\exp(\gamma - P^> - \tfrac{1}{2}P_o) + (-)^\ell \exp(-\gamma + P^< + \tfrac{1}{2}P_o)\right) \Big/ \prod_1^\ell (\gamma - \gamma_i) \\ &= \exp\left(\frac{P^< - P^>}{2}\right) \left[\exp(\gamma - \tfrac{1}{2}P) + (-)^\ell \exp(-\gamma + \tfrac{1}{2}P)\right] \Big/ \prod_1^\ell (\gamma - \gamma_i) \end{aligned} \qquad (4.34)$$

Inserting the latter expression into eqn.(4.16) and writing $R_\ell(\gamma,\zeta) = \gamma - \tfrac{1}{2}P$, we see that our patching matrix is gauge equivalent to:

$$g = \begin{pmatrix} \zeta^\ell \exp(R_\ell(\gamma)) & \dfrac{\exp(R_\ell(\gamma)) + (-)^\ell \exp(-R_\ell(\gamma))}{S(\gamma,\zeta)} \\ 0 & \zeta^{-\ell} \exp(-R_\ell(\zeta)) \end{pmatrix}, \quad S(\gamma,\zeta) = \prod_1^\ell (\gamma - \gamma_i). \qquad (4.35)$$

We may also summarise the properties of $R(\gamma)$ inferred from those of P:

$$R_\ell(\gamma_k) = n_k \frac{i\pi}{2} \, ,$$

where the n_k are a set of odd integers if ℓ is even and even integers if ℓ is odd. Thus we have

$$R_\ell(\gamma) = \frac{i\pi}{2} \sum_{k=1}^{\ell} n_k \prod_{j\neq k} \left(\frac{\gamma - \gamma_j}{\gamma_k - \gamma_j} \right), \tag{4.36}$$

from which we can deduce the coefficients $a_k(\zeta)$ which satisfy:

$$\frac{1}{2\pi i} \oint_C \frac{d\zeta}{\zeta} \zeta^m a_k(\zeta) = \delta_{1k} \delta_{mo} \tag{4.37}$$

$$\text{for } |m| \leq k-1 \text{ and } k \geq 1,$$

where C is some suitable contour enclosing the origin. The γ_k, $k = 1,\ldots,\ell$ are prescribed as before, eqn.(4.25).

In the next section we shall discuss some special cases of the general expression (4.35).

5. Special solutions

In order to gain a better understanding of the free parameters of the multi-monopole solutions implied by the patching matrix derived above, let us first of all look at the effect of rotations and translations of coordinates[31]. To do this we shall have to return to the formalism which led to the description in terms of vector bundles in the first place.

5.1 Translations

If we make a translation of coordinates by setting $x_i = x_i' + a_i$, $i = 1,2,3$ or, in the 2×2 matrix language of section 2 (eqn.(2.11)),

$$x = x' + a,$$

then the null plane equation, eqn.(2.3), tells us

$$x\pi = w \Longrightarrow x'\pi' = w' \quad \text{where} \quad \pi = \pi', \quad w = w' - a\pi'. \tag{5.1}$$

Thus, for a translation

$$\zeta = \zeta'$$

$$\gamma(\zeta) = i\left(\frac{w_2}{\pi_2} - \frac{w_1}{\pi_1}\right) = \gamma'(\zeta') - (a_1 + ia_2)\zeta' + 2a_3 + (a_1 - ia_2)\frac{1}{\zeta}\,, \tag{5.2}$$

as we would expect.

Rotations

Rotations, on the other hand, work in a more complicated way. In the 2×2 matrix language a rotation corresponds to

$$x = ax'a^{-1}, \quad a = \begin{pmatrix} \alpha & \beta \\ -\beta^* & \alpha^* \end{pmatrix} \in SU(2), \text{ if } |\alpha|^2 + |\beta|^2 = 1. \tag{5.3}$$

Thus the null plane equation (2.3) tells us that π, w transform as

$$\pi = a\pi', \quad w = aw' \tag{5.4}$$

and so,

$$\zeta = \frac{\pi_1}{\pi_2} = \frac{\alpha\zeta' + \beta}{-\beta^*\zeta' + \alpha^*} \tag{5.5}$$

while

$$\gamma(\zeta) = \gamma'(\zeta') / (\alpha + \frac{\beta}{\zeta'})(-\beta^*\zeta' + \alpha^*). \tag{5.6}$$

One result of the transformation of ζ, eqn(5.5), is that the patching matrix transforms in a non-straightforward manner. Consider a matrix of the form (2.12):

$$g = \begin{pmatrix} \zeta^{\ell} & \rho \\ 0 & \zeta^{-\ell} \end{pmatrix} = \begin{pmatrix} \left(\dfrac{\alpha\zeta' + \beta}{-\beta^*\zeta' + \alpha^*}\right)^{\ell} & \rho \\ 0 & \left(\dfrac{\alpha\zeta' + \rho}{-\beta^*\zeta' + \alpha^*}\right)^{-\ell} \end{pmatrix}.$$

This is gauge equivalent (in the sense of eqn.(2.11)) to,

$$g' = \begin{pmatrix} (\zeta')^{\ell} & \rho' \\ 0 & (\zeta')^{-\ell} \end{pmatrix}$$

where

$$\rho' = \rho(\gamma',\zeta') \,/\, (\alpha + \frac{\beta}{\zeta'})^{\ell}(-\beta^*\zeta' + \alpha^*)^{\ell} \,. \tag{5.7}$$

Using eqn.(5.7) and (5.5) we can compute the transformation law for the Laurent coefficients of ρ and ρ', the Δ's and Δ''s. We find

$$\Delta_r' = \sum_S \frac{1}{2\pi i} \oint_C \frac{d\zeta}{\zeta} \zeta^r (\alpha^* - \frac{\beta}{\zeta})^{r+\ell-1} (\alpha + \beta^*\zeta)^{\ell-1-r} \zeta^{-s} \Delta_s,$$

from which we deduce:

$$\Delta_r' = \sum_S d_{rs}^{(\ell-1)} \Delta_s, \qquad |r|,|s| \leq \ell - 1, \tag{5.8}$$

where the matrices $d_{rs}^{(\ell-1)}$ define a $(2\ell - 1)$ (non-unitary as it happens) dimensional representation of the rotation matrix related to a. The other Δ's (i.e. omitting the $2\ell - 1$ central ones in ρ) transform as some ∞ dimensional representation (inevitably non-unitary) of SU(2).

We now consider some special cases of eqn.(4.35).

5.2 $\ell = 1$

The denominator function is simply

$$S = \gamma - \gamma_1$$

where γ_1 has the form

$$\gamma_1 = a\zeta + b - \frac{a^*}{\zeta}, \qquad b = b^*, \tag{5.9}$$

corresponding precisely to a translation, as we expect. $R_1(\gamma) = \gamma - \gamma_1$ so the only possibility is a translated Prasad-Sommerfield monopole. (Rotations have no effect since Δ_0 is a scalar from eqn.(5.8) with $\ell = 1$.)

$\ell = 2$

The denominator function S is more complicated:

$$S = (\gamma-\gamma_1)(\gamma-\gamma_2) = \gamma^2 - (\gamma_1 + \gamma_2)\gamma + \gamma_1\gamma_2 . \tag{5.10}$$

Eqn.(4.25) informs us that

$$\begin{aligned} \gamma_1 + \gamma_2 &= Q_1(\zeta), \quad \text{a translation effectively,} \\ \gamma_1\gamma_2 &= Q_2(\zeta), \end{aligned} \tag{5.11}$$

and

$$\gamma_1 - \gamma_2 = \sqrt{Q_1^2 - 4Q_2}. \tag{5.12}$$

Since Q_1 is removable by a translation of coordinates we can set $\gamma_1 = \gamma_2$, $Q_1 = 0$ so that

$$\begin{aligned} \gamma_1 &= i\sqrt{Q_2(\zeta)} \\ \gamma_2 &= -i\sqrt{Q_2(\zeta)}. \end{aligned} \tag{5.13}$$

The function in the exponential, R_2, is given by

$$R_2(\gamma,\zeta) = \frac{i\pi}{2}\left[\frac{\gamma-\gamma_2}{\gamma_1 - \gamma_2} - \frac{\gamma-\gamma_1}{\gamma_2 - \gamma_1}\right] = \frac{\pi\gamma}{2\sqrt{Q_2}} \tag{5.14}$$

where we have chosen $n_1 = 1$ and $n_2 = -1$. [We shall see below this agrees with Ward's original choice, eqn.(4.9).]

Because of the form of eqn.(5.14) we can perform a rotation of coordinates to simplify Q_2. It is fairly easy to see that we can use a rotation to set the coefficient of ζ^2 (and $\frac{1}{\zeta^2}$) in Q_2 to zero, and to make the coefficient of ζ (and $\frac{1}{\zeta}$) real. In other words $Q_2(\zeta)$ can be reduced to a factored form,

$$Q_2(\zeta) = h\left(\frac{1}{\zeta} + k\right)(\zeta - k), \quad h \text{ and } k \text{ real.} \tag{5.15}$$

From eqn.(4.37) there is just one constraint to satisfy namely,

$$\frac{1}{2\pi i}\oint_C \frac{d\zeta}{\zeta}\frac{\pi}{2\sqrt{Q_2}} = 1,$$

or

$$\sqrt{h} = \frac{1}{4i}\oint\frac{d\zeta}{\zeta}\frac{1}{\sqrt{(\frac{1}{\zeta} + k)(\zeta - k)}} \tag{5.16}$$

where the contour C encloses the cut extending from $\zeta = 0$ to $\zeta = k$ but excludes the cut from $\zeta = -\frac{1}{k}$ to ∞. Eqn.(5.16) defines h as a function of k (a complete elliptic integral of the first kind) and k is the only free parameter left over. Note that when $k = 0$, $\sqrt{h} = \frac{\pi}{2}$ and the function in the patching matrix collapses - precisely to the one given in eqn.(4.9). For $k \neq 0$, the patching matrix is essentially the one given by Ward[29] (except that he used a parameter p $(= \frac{2k}{1-k^2})$ instead of k)

and $$\sqrt{h} = \frac{\pi}{2} \sum_{0}^{\infty} (-)^r k^{2r} \begin{pmatrix} -\frac{1}{2} \\ r \end{pmatrix}^2, \quad \text{if} \quad |k| < 1. \tag{5.17}$$

Ward also showed by calculating to lowest order in p and the coordinates $\underline{x}$ that the Higgs' field has a pair of zeros lying on the x_1 axis symmetrically about the origin and displaced from each other by an amount proportional to p. If p (and not k) is the actual distance between finitely displaced monopoles then it would have to be that the parameter k is restricted to be less than 1. For other solutions there may be other inequalities to be satisfied by the parameters of the polynomials $Q_k(\zeta)$ but it is not really clear how to tackle this problem.

Finally we remark that, choosing each γ_i, $i = 1,\ldots,\ell$ to be a multiple of $\frac{i\pi}{2}$ we can recover the special cases described before in eqn.(4.14),(4.15). Regarding the more general solutions as lying in a neighbourhood of these special ones we are tempted to suppose that the integers n_k, $k = 1,\ldots,\ell$ will in fact be as small as they can be compatible with being distinct.

6. Conclusion

It is difficult to see how to proceed further along the lines described above because the explicit construction of solutions involves the solution of polynomial equations of arbitrary degree. Moreover the constraints occur in terms of transcendental functions defined by contour integrals of which eqn.(5.16) is the simplest example.

On the other hand, optimistically, it may be possible to recreate the whole structure for bigger gauge groups and see the generalisation of the work described, for example, in ref. (5). It may also be possible to understand the relationship between the several ways of looking at monopoles - via ADHM(17), Bäcklund transformations(8,30) and the work described in these lectures. There is clearly much to be done and only time will tell whether this particular brand of four-dimensional soliton is a true hint of integrability in four dimensions.

Added Note

A recent review article by O'Raifeartaigh and Rouhani(32) may be found helpful also.

Acknowledgements

I would like to thank Claus Montonen and Jarmo Hieterinta for their kind invitation and the Research Institute for Theoretical Physics of the University of Helsinki for its hospitality. I am also grateful to David Fairlie and Peter Goddard for many discussions about monopoles.

References

1. A.C.Scott, F.Y.F.Chiu and D.W.McLaughlin, Proceedings of the IEEE Vol. 61, No. 10 (1973) 1443.
2. G. 't Hooft, Nucl. Phys. B 79 (1974 276.
   A.M.Polyakov, JETP Lett. 20 (1974) 194
3. For a review see P. Goddard and D. Olive, Reports on Progress in Physics 41 (1978) 1357.
4. E.B.Bogomolny, Sov. J. Nucl. Phys. 24 (1976) 449.
   S. Coleman, S. Parke, A. Neveu and C.M.Sommerfield, Phys. Rev. D15 (1977) 544.
5. For a review of this see D. Olive, 'Classical solutions in gauge theories - spherically symmetric monopoles - Lax pairs and Toda Lattices', lectures given at the International Summer Institute on Theoretical Physics, Bad Honnef. September 1980.
6. N. Manton, Nucl. Phys. B126 (1977) 525.
   L. O'Raifeartaigh, S. Y.Park , K.C.Wali, Phys. Rev. 20D (1979) 1941.
7. C. Taubes, 'Existence of multi-monopole solutions' to appear in Comm. Math. Phys.
   A. Jaffe and C. Taubes, Vortices and Monopoles (Birkhauser, Boston 1980).
8. P. Forgacs, Z. Horvath and L. Palla, Phys. Lett. 99B (1981) 232.
9. R.S.Ward, 'A Yang-Mills-Higgs monopole of charge 2' to appear in Comm. Math. Phys.
10. F. Ernst, Phys. Rev. 167 (1968) 1175.
11. B.K.Harrison, Phys. Rev. Lett. 41 (1978) 1197.
    G. Neugebauer, J. Phys. A12 (1979) L67.
12. M.K.Prasad and C.M.Sommerfield, Phys. Rev. Lett. 35 (1975) 760.
13. E.Corrigan, D.B.Fairlie, J. Nuyts and D. Olive, Nucl. Phys. B106 (1976) 475.

14. E.B.Bogomolny, Sov. J. Nucl. Phys. 24 (1976) 449.

15. For reviews see for example

D. Olive, Rivista del Nuovo Cimento 2 (1979) 1.

E. Corrigan, Phys. Reps. 49C (1979) 95.

M.F.Atiyah, Geometry of Yang-Mills Fields, Lezione Fermioni, Pisa 1979.

E. Corrigan and P. Goddard, Lecture notes in Physics 129, Geometrical and Topological Methods in Gauge Theories, Eds. J.P.Harnad and S. Shnider (Springer Verlag 1980).

16. M.F.Atiyah, N.J.Hitchin, V.G.Drinfeld and Yu I. Manin, Phys. Letts. 65A (1978) 185.

17. W. Nahm, Phys. Letts. 90B (1980) 413, 93B (1980) 42.

18. R.S.Ward, Phys. Letts. 61A (1977) 81.

19. M.F.Atiyah and R.S.Ward, Comm. Math. Phys. 55 (1977) 117.

20. F. Wilczek, Quark Confinement and Field Theory, eds. D. Stump and D. Weingarten (John Wiley and Sons, New York (1977)).

E. Corrigan and D.B.Fairlie, Phys. Letts. 67B (1977) 69.

R. Jackiw, C. Nohl and C. Rebbi, Phys. Rev. D15 (1977) 1642.

21. N. Manton, Nucl. Phys. B135 (1978) 319.

22. M.K.Prasad and P. Rossi, MIT Preprint CTP 903 (1980).

23. E. Weinberg, Phys. Rev. D20 (1979) 936.

24. C.N.Yang, Phys. Rev. Letts. 38 (1977) 1377.

25. E. Corrigan, D.B.Fairlie, P. Goddard and R. Yates, Comm. Math. Phys. 58 (1978) 2528.

26. A useful review of this section is in M.K.Prasad, Physica 1D (1980) 167.

27. M.A.Lohe, Nucl. Phys. B142 (1978) 236.

D.J.Bruce, Nucl. Phys. B142 (1978) 253.

28. M.K.Prasad, 'Exact Yang-Mills-Higgs Monopole solutions of arbitrary topological charge', Comm. Math. Phys. to be published.

29. R.S.Ward, 'Two Yang-Mills-Higgs monopoles close together' Dublin preprint, March 1981.

30. P. Forgacs, Z. Horvath and L. Palla, 'Non-linear superposition of monopoles' March 1981.

31. E. Corrigan and P. Goddard, 'An n monopole solution with 4n - 1 degrees of freedom' DAMTP 81/9 March 1981.

32. L. O'Raifeartaigh and S. Rouhani, Schladming lectures (1981), Dublin preprint DIAS-STP-81-03.

Lecture Notes in Physics

Vol. 114: Stellar Turbulence. Proceedings, 1979. Edited by D. F. Gray and J. L. Linsky. IX, 308 pages. 1980.

Vol. 115: Modern Trends in the Theory of Condensed Matter. Proceedings, 1979. Edited by A. Pekalski and J. A. Przystawa. IX, 597 pages. 1980.

Vol. 116: Mathematical Problems in Theoretical Physics. Proceedings, 1979. Edited by K. Osterwalder. VIII, 412 pages. 1980.

Vol. 117: Deep-Inelastic and Fusion Reactions with Heavy Ions. Proceedings, 1979. Edited by W. von Oertzen. XIII, 394 pages. 1980.

Vol. 118: Quantum Chromodynamics. Proceedings, 1979. Edited by J. L. Alonso and R. Tarrach. IX, 424 pages. 1980.

Vol. 119: Nuclear Spectroscopy. Proceedings, 1979. Edited by G. F. Bertsch and D. Kurath. VII, 250 pages. 1980.

Vol. 120: Nonlinear Evolution Equations and Dynamical Systems. Proceedings, 1979. Edited by M. Boiti, F. Pempinelli and G. Soliani. VI, 368 pages. 1980.

Vol. 121: F. W. Wiegel, Fluid Flow Through Porous Macromolecular Systems. V, 102 pages. 1980.

Vol. 122: New Developments in Semiconductor Physics. Proceedings, 1979. Edited by F. Beleznay et al. V, 276 pages. 1980.

Vol. 123: D. H. Mayer, The Ruelle-Araki Transfer Operator in Classical Statistical Mechanics. VIII, 154 pages. 1980.

Vol. 124: Gravitational Radiation, Collapsed Objects and Exact Solutions. Proceedings, 1979. Edited by C. Edwards. VI, 487 pages. 1980.

Vol. 125: Nonradial and Nonlinear Stellar Pulsation. Proceedings, 1980. Edited by H. A. Hill and W. A. Dziembowski. VIII, 497 pages. 1980.

Vol. 126: Complex Analysis, Microlocal Calculus and Relativistic Quantum Theory. Proceedings, 1979. Edited by D. Iagolnitzer. VIII, 502 pages. 1980.

Vol. 127: E. Sanchez-Palencia, Non-Homogeneous Media and Vibration Theory. IX, 398 pages. 1980.

Vol. 128: Neutron Spin Echo. Proceedings, 1979. Edited by F. Mezei. VI, 253 pages. 1980.

Vol. 129: Geometrical and Topological Methods in Gauge Theories. Proceedings, 1979. Edited by J. Harnad and S. Shnider. VIII, 155 pages. 1980.

Vol. 130: Mathematical Methods and Applications of Scattering Theory. Proceedings, 1979. Edited by J. A. DeSanto, A. W. Sáenz and W. W. Zachary. XIII, 331 pages. 1980.

Vol. 131: H. C. Fogedby, Theoretical Aspects of Mainly Low Dimensional Magnetic Systems. XI, 163 pages. 1980.

Vol. 132: Systems Far from Equilibrium. Proceedings, 1980. Edited by L. Garrido. XV, 403 pages. 1980.

Vol. 133: Narrow Gap Semiconductors Physics and Applications. Proceedings, 1979. Edited by W. Zawadzki. X, 572 pages. 1980.

Vol. 134: $\gamma\gamma$ Collisions. Proceedings, 1980. Edited by G. Cochard and P. Kessler. XIII, 400 pages. 1980.

Vol. 135: Group Theoretical Methods in Physics. Proceedings, 1980. Edited by K. B. Wolf. XXVI, 629 pages. 1980.

Vol. 136: The Role of Coherent Structures in Modelling Turbulence and Mixing. Proceedings 1980. Edited by J. Jimenez. XIII, 393 pages. 1981.

Vol. 137: From Collective States to Quarks in Nuclei. Edited by H. Arenhövel and A. M. Saruis. VII, 414 pages. 1981.

Vol. 138: The Many-Body Problem. Proceedings 1980. Edited by R. Guardiola and J. Ros. V, 374 pages. 1981.

Vol. 139: H. D. Doebner, Differential Geometric Methods in Mathematical Physics. Proceedings 1981. VII, 329 pages. 1981.

Vol. 140: P. Kramer, M. Saraceno, Geometry of the Time-Dependent Variational Principle in Quantum Mechanics. IV, 98 pages. 1981.

Vol. 141: Seventh International Conference on Numerical Methods in Fluid Dynamics. Proceedings. Edited by W. C. Reynolds and R. W. MacCormack. VIII, 485 pages. 1981.

Vol. 142: Recent Progress in Many-Body Theories. Proceedings. Edited by J. G. Zabolitzky, M. de Llano, M. Fortes and J. W. Clark. VIII, 479 pages. 1981.

Vol. 143: Present Status and Aims of Quantum Electrodynamics. Proceedings, 1980. Edited by G. Gräff, E. Klempt and G. Werth. VI, 302 pages. 1981.

Vol. 144: Topics in Nuclear Physics I. A Comprehensive Review of Recent Developments. Edited by T.T.S. Kuo and S.S.M. Wong. XX, 567 pages. 1981.

Vol. 145: Topics in Nuclear Physics II. A Comprehensive Review of Recent Developments. Proceedings 1980/81. Edited by T. T. S. Kuo and S. S. M. Wong. VIII, 571-1.082 pages. 1981.

Vol. 146: B. J. West, On the Simpler Aspects of Nonlinear Fluctuating. Deep Gravity Waves. VI, 341 pages. 1981.

Vol. 147: J. Messer, Temperature Dependent Thomas-Fermi Theory. IX, 131 pages. 1981.

Vol. 148: Advances in Fluid Mechanics. Proceedings, 1980. Edited by E. Krause. VII, 361 pages. 1981.

Vol. 149: Disordered Systems and Localization. Proceedings, 1981. Edited by C. Castellani, C. Castro, and L. Peliti. XII, 308 pages. 1981.

Vol. 150: N. Straumann, Allgemeine Relativitätstheorie und relativistische Astrophysik. VII, 418 Seiten. 1981.

Vol. 151: Integrable Quantum Field Theory. Proceedings, 1981. Edited by J. Hietarinta and C. Montonen. V, 251 pages. 1982.

Selected Issues from

Lecture Notes in Mathematics

Vol. 728: Non-Commutative Harmonic Analysis. Proceedings, 1978. Edited by J. Carmona and M. Vergne. V, 244 pages. 1979.

Vol. 729: Ergodic Theory. Proceedings 1978. Edited by M. Denker and K. Jacobs. XII, 209 pages. 1979.

Vol. 730: Functional Differential Equations and Approximation of Fixed Points. Proceedings, 1978. Edited by H.-O. Peitgen and H.-O. Walther. XV, 503 pages. 1979.

Vol. 731: Y. Nakagami and M. Takesaki, Duality for Crossed Products of von Neumann Algebras. IX, 139 pages. 1979.

Vol. 733: F. Bloom, Modern Differential Geometric Techniques in the Theory of Continuous Distributions of Dislocations. XII, 206 pages. 1979.

Vol. 735: B. Aupetit, Propriétés Spectrales des Algèbres de Banach. XII, 192 pages. 1979.

Vol. 738: P. E. Conner, Differentiable Periodic Maps. 2nd edition, IV, 181 pages. 1979.

Vol. 742: K. Clancey, Seminormal Operators. VII, 125 pages. 1979.

Vol. 755: Global Analysis. Proceedings, 1978. Edited by M. Grmela and J. E. Marsden. VII, 377 pages. 1979.

Vol. 756: H. O. Cordes, Elliptic Pseudo-Differential Operators – An Abstract Theory. IX, 331 pages. 1979.

Vol. 760: H.-O. Georgii, Canonical Gibbs Measures. VIII, 190 pages. 1979.

Vol. 762: D. H. Sattinger, Group Theoretic Methods in Bifurcation Theory. V, 241 pages. 1979.

Vol. 765: Padé Approximation and its Applications. Proceedings, 1979. Edited by L. Wuytack. VI, 392 pages. 1979.

Vol. 766: T. tom Dieck, Transformation Groups and Representation Theory. VIII, 309 pages. 1979.

Vol. 771: Approximation Methods for Navier-Stokes Problems. Proceedings, 1979. Edited by R. Rautmann. XVI, 581 pages. 1980.

Vol. 773: Numerical Analysis. Proceedings, 1979. Edited by G. A. Watson. X, 184 pages. 1980.

Vol. 775: Geometric Methods in Mathematical Physics. Proceedings, 1979. Edited by G. Kaiser and J. E. Marsden. VII, 257 pages. 1980.

Vol. 779: Euclidean Harmonic Analysis. Proceedings, 1979. Edited by J. J. Benedetto. III, 177 pages. 1980.

Vol. 780: L. Schwartz, Semi-Martingales sur des Variétés, et Martingales Conformes sur des Variétés Analytiques Complexes. XV, 132 pages. 1980.

Vol. 782: Bifurcation and Nonlinear Eigenvalue Problems. Proceedings, 1978. Edited by C. Bardos, J. M. Lasry and M. Schatzman. VIII, 296 pages. 1980.

Vol. 783: A. Dinghas, Wertverteilung meromorpher Funktionen in ein- und mehrfach zusammenhängenden Gebieten. Edited by R. Nevanlinna and C. Andreian Cazacu. XIII, 145 pages. 1980.

Vol. 786: I. J. Maddox, Infinite Matrices of Operators. V, 122 pages. 1980.

Vol. 787: Potential Theory, Copenhagen 1979. Proceedings, 1979. Edited by C. Berg, G. Forst and B. Fuglede. VIII, 319 pages. 1980.

Vol. 791: K. W. Bauer and S. Ruscheweyh, Differential Operators for Partial Differential Equations and Function Theoretic Applications. V, 258 pages. 1980.

Vol. 792: Geometry and Differential Geometry. Proceedings, 1979. Edited by R. Artzy and I. Vaisman. VI, 443 pages. 1980.

Vol. 793: J. Renault, A Groupoid Approach to C*-Algebras. III, 160 pages. 1980.

Vol. 798: Analytic Functions, Kozubnik 1979. Proceedings. Edited by J. Ławrynowicz. X, 476 pages. 1980.

Vol. 799: Functional Differential Equations and Bifurcation. Proceedings 1979. Edited by A. F. Izé. XXII, 409 pages. 1980.

Vol. 801: K. Floret, Weakly Compact Sets. VII, 123 pages. 1980.

Vol. 802: J. Bair, R. Fourneau, Etude Géometrique des Espaces Vectoriels II. VII, 283 pages. 1980.

Vol. 804: M. Matsuda, First Order Algebraic Differential Equations. VII, 111 pages. 1980.

Vol. 805: O. Kowalski, Generalized Symmetric Spaces. XII, 187 pages. 1980.

Vol. 807: Fonctions de Plusieurs Variables Complexes IV. Proceedings, 1979. Edited by F. Norguet. IX, 198 pages. 1980.

Vol. 809: I. Gumowski and Ch. Mira, Recurrences and Discrete Dynamic Systems. VI, 272 pages. 1980.

Vol. 810: Geometrical Approaches to Differential Equations. Proceedings 1979. Edited by R. Martini. VII, 339 pages. 1980.

Vol. 816: L. Stoica, Local Operators and Markov Processes. VIII, 104 pages. 1980.

Vol. 819: Global Theory of Dynamical Systems. Proceedings, 1979. Edited by Z. Nitecki and C. Robinson. IX, 499 pages. 1980.

Vol. 821: Statistique non Paramétrique Asymptotique. Proceedings, 1979. Edited by J.-P. Raoult. VII, 175 pages. 1980.

Vol. 826: Ph. G. Ciarlet et P. Rabier, Les Equations de von Kármán. VI, 181 pages. 1980.

Vol. 827: Ordinary and Partial Differential Equations. Proceedings, 1978. Edited by W. N. Everitt. XVI, 271 pages. 1980.

Vol. 831: Representation Theory I. Proceedings, 1979. Edited by V. Dlab and P. Gabriel. XIV, 373 pages. 1980.

Vol. 832: Representation Theory II. Proceedings, 1979. Edited by V. Dlab and P. Gabriel. XIV, 673 pages. 1980.

Vol. 836: Differential Geometrical Methods in Mathematical Physics. Proceedings, 1979. Edited by P. L. García, A. Pérez-Rendón, and J. M. Souriau. XII, 538 pages. 1980.

Vol. 837: J. Meixner, F. W. Schäfke and G. Wolf, Mathieu Functions and Spheroidal Functions and their Mathematical Foundations Further Studies. VII, 126 pages. 1980.

Vol. 838: Global Differential Geometry and Global Analysis. Proceedings 1979. Edited by D. Ferus et al. XI, 299 pages. 1981.

Vol. 840: D. Henry, Geometric Theory of Semilinear Parabolic Equations. IV, 348 pages. 1981.

Vol. 841: A. Haraux, Nonlinear Evolution Equations- Global Behaviour of Solutions. XII, 313 pages. 1981.

Vol. 842: Séminaire Bourbaki vol. 1979/80. Exposés 543–560. IV, 317 pages. 1981.

Vol. 843: Functional Analysis, Holomorphy, and Approximation Theory. Proceedings. Edited by S. Machado. VI, 636 pages. 1981.

Vol. 845: A. Tannenbaum, Invariance and System Theory: Algebraic and Geometric Aspects. X, 161 pages. 1981.

Vol. 849: P. Major, Multiple Wiener-Itô Integrals. VII, 127 pages. 1981.